Tourism and the Spectre of Unlimited Change

This insightful volume forms a sequel to *Living with Tourism: Negotiating Identities in a Turkish Village*, tracking the tourism development and associated social change in the small town of Göreme, in Turkey's Cappadocia region, within the last two decades.

Carefully crafted chapters explore the significant changes in the tourism forms, place identity, and social relations in the town. On one level, tourism business and Göreme's 'living with tourism' has matured and thrived: the place has, due largely to its booming hot-air ballooning sector, become an 'Instagram sensation'; some Göreme families have become very wealthy; and tourism has enabled many local women, as well as men, to 'craft new selves'. On another level, new inequalities and tensions constantly emerge: some families remain poor; gentrification and hotel developments in the older 'cave-house' neighbourhoods have led to the disintegration of community; and many people, including those who are now wealthy, talk often with a sense of nostalgia and regret about what Göreme has become. This book is a groundbreaking longitudinal account, recounting the story of the place and people of Göreme 'still living with tourism' after 40 years, showing how broader contemporary tourism trends, such as changes in tourism markets and use of digital technology, and increased security fears, manifest at the local level in tourism destinations.

This book provides new insights for scholars of tourism, anthropology, geography, and social studies, who wish to gain a deeper understanding of this global phenomenon in the contemporary world.

Hazel Tucker has a Ph.D. in Social Anthropology from the University of Durham and is Professor of Tourism at the University of Otago, New Zealand. Specialising in the area of tourism's influences on socio-cultural relationships and change, Professor Tucker publishes also on gender and tourism, colonialism/postcolonialism, and emotional dimensions of tourism encounters. She serves as Associate Editor of *Annals of Tourism Research* and is an associate at Equality in Tourism.

Contemporary Geographies of Leisure, Tourism and Mobility

Series Editor: C. Michael Hall,
Professor at the Department of Management, College of Business and Economics, University of Canterbury, Christchurch, New Zealand

The aim of this series is to explore and communicate the intersections and relationships between leisure, tourism and human mobility within the social sciences.

It will incorporate both traditional and new perspectives on leisure and tourism from contemporary geography, e.g. notions of identity, representation and culture, while also providing for perspectives from cognate areas such as anthropology, cultural studies, gastronomy and food studies, marketing, policy studies and political economy, regional and urban planning, and sociology, within the development of an integrated field of leisure and tourism studies.

Also, increasingly, tourism and leisure are regarded as steps in a continuum of human mobility. Inclusion of mobility in the series offers the prospect to examine the relationship between tourism and migration, the sojourner, educational travel, and second home and retirement travel phenomena.

The series comprises two strands:

Contemporary Geographies of Leisure, Tourism and Mobility aims to address the needs of students and academics, and the titles will be published simultaneously in hardback and paperback. Titles include:

Second Homes and Climate Change
Bailey Ashton Adie and C. Michael Hall

Gen Z, Tourism, and Sustainable Consumption
The Most Sustainable Generation Ever?
Edited by Siamak Seyfi, C. Michael Hall and Marianna Strzelecka

Tourism and the Spectre of Unlimited Change
Living with Tourism in a Turkish Village Revisited
Hazel Tucker

For more information about this series, please visit: www.routledge.com/Contemporary-Geographies-of-Leisure-Tourism-and-Mobility/book-series/SE0522

Tourism and the Spectre of Unlimited Change

Living with Tourism in a Turkish Village Revisited

Hazel Tucker

Routledge
Taylor & Francis Group

LONDON AND NEW YORK

First published 2024
by Routledge
4 Park Square, Milton Park, Abingdon, Oxon OX14 4RN

and by Routledge
605 Third Avenue, New York, NY 10158

Routledge is an imprint of the Taylor & Francis Group, an informa business

© 2024 Hazel Tucker

British Library Cataloguing-in-Publication Data
A catalogue record for this book is available from the British Library

ISBN: 978-0-367-42957-7 (hbk)
ISBN: 978-1-032-32428-9 (pbk)
ISBN: 978-1-003-01120-0 (ebk)

DOI: 10.4324/9781003011200

Typeset in Galliard
by codeMantra

Contents

Figures

Acknowledgements

I owe enormous thanks to the people of Göreme who have supported me in my research endeavours with their friendship, guidance, and generosity for over three decades. My heartfelt thanks go especially to Abbas and Senem Köse, whose kind hospitality and openness have been invaluable. I enjoy our time together immensely and learn so much from our conversations. I am indebted also to Mehmet Bozlak (always good for a story or two), Neriman Güngör (always good for a delicious kuru fasulye), Ruth Lockwood, Mustafa Çiftçi, Andus Emge, Mehmet Daşdeler, and Ali Yavuz for their generous practical support and friendship. Your generosity has made my repeated 'returns to the field' both comfortable and enjoyable.

My immense gratitude goes, too, to all the other people who have helped me in my research by sharing your perspectives on 'living with tourism', including but certainly not limited to: Osman Atak, Mustafa Durmaz, Fatma Şafak, 'Walking Mehmet', Hayriye and Hanife, Emine, Zekiye, and Fatma – the wonderfully strong women with their various jewellery shops and stalls, 'Bicycle Mustafa', Mustafa Demirci, Dawn Harris, Leyla, Hanife and Emre Köse, Lars Eric Möre, Hava and Şerife, Jeyda Lockwood, Jessica Lee, Sarah Cox, Sabina, Şenol, Fatih, the Tekkaya family, Halil, Pembe, Emine, Zubeyde, Idris, Refik, Numan and Hatice Atak, and so the list goes on. Thank you for our interesting conversations, for patiently answering my questions, and for sharing your stories with me for this book.

Many thanks to Liam Scaife for preparing the map and to Animesh Tripathi for his technical assistance. Thanks go also to the Göreme Tourism Development Cooperative for providing old Göreme photographs, as indicated in the book, and to Mustafa Demirci for providing the Aydınlı Cave-Hotel photograph. The remaining photographs are my own.

I extend my warmest thanks to my amazing colleagues and friends Dr Stu Hayes, Professor Tamara Kohn, and Dr Jasmine Zhang for reading and commenting on chapters and sections of the manuscript, and to Pat Hamlin whose positivity, insightful comments, and questions while reading through the manuscript from beginning to end were invaluable.

I am immensely grateful to have received a University of Otago Writing Grant which allowed me time to work on this manuscript. Additionally, the

University of Otago has enabled my repeated 'returns to the field' during the past two decades by providing periodic breaks to travel from New Zealand to Turkey to conduct fieldwork, and frequently providing funds for this travel. Without those regular opportunities to spend time in Göreme, this book would not have been possible.

I owe thanks to Harriet Cunningham at Routledge for her enduring patience and continuous positivity and kindness when I asked for yet another extension.

In the Department of Tourism at Otago, and Dunedin more generally, I have greatly appreciated the company and support of my colleagues and friends for getting me through the thick treacle of Covid-19 while simultaneously I worked through the thick treacle of writing this book. I immensely enjoy our conversations, the food and the wines, the barbecues, the camping trips, the swims in the sea and the lakes, the putting the world to rights... Most of all, I appreciate the laughter.

As always, to my parents Carole and Jim Tucker, whose love and encouragement have boosted my life journey at every stage. And to Rob and Liam, the two big loves in my life. Liam, your commitment to creating a better world is astoundingly wonderful – please never give up. I should not forget to thank Poppy for taking me on wonderful runs and walks at the end of each day of writing.

In loving memory of Kaili Kidner, who forever holds a special place in Göreme's story and in Göremeli hearts.

1 Introduction

During the summer of 2019, on a Friday afternoon as the men were coming out of the mosque, I met Mustafa, the director of the Göreme Tourism Development Cooperative. After enquiring about each other's health, I told Mustafa that I was doing another book about Göreme, a follow-up of the previous book, to write about the ongoing changes during the 20 years or so since I wrote that first book. In response, Mustafa said that he wished Göreme had stayed as it was 20 years ago; "It was good then", he said, "Now it's like a runaway car, a car without brakes and we don't know where the car will end up going, because it's out of control and we've got no way of stopping it". "Çok üzgünüm" (I'm *very* sad), he said, "we want to press on the brakes, but there are no brakes and we don't know how to find them". "Çok üzgünüm", he repeated.

I consider what I was hearing from Mustafa to be the spectre of unlimited change. My use of the word *unlimited*, as I will explain in more detail later, is the foreboding of change, the speed and reach of which appears not only to be without bounds but also beyond any particular, or any*one's*, intention or determination. During that summer in 2019, many Göreme people expressed sentiments similar to Mustafa's, that tourism – along with life in general – was hurtling out of control at such speed that there was seemingly no way to stop it, or even to slow it down, nor to steer it in a particular direction. Mustafa's words resonate with what Anthony Giddens (2003) calls the 'runaway world', and with what Thomas Eriksen describes as the 'overheating' characteristic of the contemporary neoliberal world; 'the confluence of several runaway processes, forms of growth that were meaningful and purposive for a long time before reaching a point where the unintended side-effects were threatening to become more noticeable than the intentional or functional effects' (2016, p. 22). Indeed, Mustafa's sentiments conjure a looming threat – of change without intention and without limits: change which might make every bit of life and living unrecognisable from what had been before. His words conjure a looming spectre of unrelenting indeterminacy, or – as Anna Tsing (2015) has put it – of 'living without those handrails which once made us think we knew, collectively, where we were going' (p. 2).

DOI: 10.4324/9781003011200-1

My previous book, which I will from now on refer to as *Living with Tourism (1)* (Tucker, 2003), portrayed tourism as in many respects providing those handrails, since living with tourism as it was then did yield a level of relative economic stability that the people of Göreme had long hoped for, as well as a concomitant sense of determinacy over their lives. Prior to gaining that stability through actively making their living through tourism, most Göreme families had struggled to eke out a living through their subsistence farming practices supplemented by labouring or truck driving work, or outward migration of some family members. Indeed, even with the relative economic stability which tourism provided, during my first phases of ethnographic research in the mid-to-late 1990s, I still sometimes heard people voicing their fears about whether or not they would be able to feed themselves throughout the coming winter months. There was also then still a sense in Göreme of time as cyclical, as well as of work as *re*productive. In *Living with Tourism (1)* I wrote about how the frequently uttered rhyme "*Yazın, turizm – kışın, kuru üzüm*" (In summer, tourism – in winter, dried grapes) portrayed both the cyclic rhythm of each year that living with tourism engendered and the dualism of life at that time. The handrails of stability were formed not only out of the establishment of locally owned tourism business but also by *continuity*; continuity of customary farming practices and of the boundaries of moral propriety according to codes of honour and shame. *Living with Tourism (1)*, which provides an in-depth account of both change and continuity during Göreme's first two decades (1980s–90s) of living with tourism, thus portrayed an overall picture of Göreme as being an example of successful community-led tourism development.

This book, as a sequel of sorts, picks up where *Living with Tourism (1)* left off, although many of the chapters refer back to the 1980s–90s as the back-story to what is being discussed in relation to the period 2000–19. My original intentions of making these two decades the main focus of this book (#2) continue to seem appropriate, since the onset of the Covid-19 pandemic in early 2020 provides a suitable book-end, both metaphorically and literally. Focusing on the period 2000–19 allows me to tell the story, or stories, of tourism happenings in Göreme which led up to the sense of foreboding expressed by Mustafa and others, the sense that the handrails of determinacy and familiarity were failing. On one level, tourism business and Göreme's 'living with tourism' could be said to have matured and flourished during this period; by 2019, largely due to its thriving hot-air ballooning sector with over 100 balloons flying each morning, coupled with the rise of social media during that decade, Göreme had become an *Instagram* sensation. With over 200 hotels and 120 restaurants, plus numerous souvenir shops and tourism activity businesses operating in the village, the majority of which were locally owned, many Göreme families had become very wealthy indeed. On another level, new inequalities and tensions had emerged; hotel developments and gentrification in the old cave-house neighbourhoods had, in many respects, led to displacement and disintegration of the community. Rather than viewing these happenings as resulting from malevolent or otherwise intentions, however, I have observed in Göreme how, as Eriksen (2016, p. viii) puts it, 'changes may take unexpected direction, which were neither foreseen nor desired at the outset'.

Hence, the various multilayered and undulating happenings at play manifested as deep contradictions and ambivalences in the everyday experiences of Göreme life. Significant in this was that, during the two decades from 2000 to 2019, not only did Göreme's new-found global fame come to attract a tourist clientele which was altogether different from the backpacker tourists who had come to Göreme in the 1980s and 1990s, but the new tourist clientele came to be seen as plentiful, perhaps even grow-able, in number. Hence, as I will go onto explain, tourism – and everything it might bring – went from being seen as a *limited good* to something with *unlimited* potential. For this reason, Göreme's tourism attracted the attention of an increasing number and variety of external players who saw it as a resource from which unlimited good – or profit – could be extracted. Through storying the dynamic undulations of these changes, this book attends to the everyday experiences, practices, and ambiguities which come with the gradual switching from a moral economy – based on an image of limited good and constituted of reciprocity and community ties, to a prospect of unlimited good – and with it, I argue, the spectre of unlimited, unbounded change.

The back-story; *Living with Tourism (1)*

I first visited the Cappadocia region of central Turkey in 1984, as a teenage backpacker rather than ethnographer. As there was not much tourism activity or business there at that time, for accommodation, my friends and I put our little tent up on a piece of scrubland in the middle of Göreme, then a dusty village of honey-coloured cave-house neighbourhoods set amongst a moon-like landscape of giant rock cones called "fairy-chimneys". I visited Göreme again a few times during the late 1980s, when a number of small-scale backpacker accommodation, restaurant, and activity businesses were starting up. By the time I returned to Göreme to conduct ethnographic research for my anthropology postgraduate degree during the mid-to-late 1990s, the village had become more fully established as a destination for backpacker tourists, with 60 or so locally owned guest houses and around 15 small cafés and restaurants. My doctoral thesis (1999), entitled *Living with Tourism: Negotiating Identities in a Turkish Village*, comprised an in-depth ethnographic portrayal of those first two decades of tourism development and social change in Göreme, and was subsequently published as a monograph in 2003 under the same title. While reading *Living with Tourism (1)* would be the ideal way for readers of this second book to acquire a fuller back-story understanding, I will summarise some of that first book's key points here by way of introduction.

The close proximity of Göreme village to the Göreme Open-Air Museum, a major cultural tourism attraction of the region and listed as a UNESCO World Heritage Site in 1985, meant that the village itself became incorporated into the Göreme National Park upon the park's formation that same year. This inclusion of the village within the national park boundary went on to have significant implications not only in relation to how Göreme developed as a tourist destination during the 1980s and 1990s but also regarding the particular ways in which tourism became incorporated into the daily lives of villagers. There were two key ways, as discussed in *Living with Tourism (1)* (see especially pages 10–14; 34–36; 160–166), in which the national park formation affected these particular developments. Firstly, the formation of the Göreme National Park expressly included the "picturesque village life" existing within the boundary of the park as an important element, or resource, of the park. This picturesque village life was thus deemed worthy of protection and preservation and, since that life was lived primarily in Göreme's old cave-house neighbourhoods, those neighbourhoods were placed under preservation zone restrictions, with any alterations to houses or rock structures (fairy chimneys) in those areas requiring permission.

Consequently, as tourism developed, a spatial and social dualism arose, which I referred to throughout *Living with Tourism (1)*. Spatially, a tourism realm developed which was located in the central areas of Göreme village as well as in the *pansiyons*, or guest houses, while the non-tourism realm comprised the winding residential streets and houses of the old neighbourhoods where villagers continued to carry out their daily lives. Socially, this dualism manifested most profoundly as a gendered separation of lives and roles which, while borne out of a strict gender segregation already inherent in Göreme's Muslim society, was strengthened through tourism development processes during the 1980s–90s. During those first two decades of Göreme's tourism, while men became tourism entrepreneurs – setting up and running tourism businesses, women's lives continued to be centred in and around the household since it was deemed inappropriate for women to work in tourism or to be present in touristic areas. As I summarised in the Conclusion chapter of *Living with Tourism (1)* (p. 182), 'Juxtaposed with the gendered separation of lives and roles in the village, the tensions seen to exist between 'tradition' and 'tourism' have created the emergence of two distinct realms in Göreme for both tourists and villagers'.

The second major way in which Göreme's positioning within the national park boundary influenced the particular ways in which tourism developed related to the national park protection laws placing restrictions on large-scale construction. As the Cappadocia region became a centre for Turkish 'cultural tourism development' during the 1980s, the national government's 1982 Tourism Encouragement Act legislation put into place generous incentives for private tourism investments while annulling previous prohibition of foreign companies acquiring real estate. Hence, national and international hotel corporations moved into the area developing large-scale tourism facilities to accommodate tour groups. Having to be placed outside of the national park perimeter, however, these large-scale tourism developments tended to occur in the towns of Ürgüp, Avanos, and Nevşehir, each approximately ten kilometres from Göreme village. Meanwhile, due to its 'protected' status within the national park, Göreme remained relatively unaffected by the 'mass' forms of tourism developing outside of the park boundary. Rather, tourism in Göreme was low on capital investment and developed in a pattern of locally owned small and micro-businesses which were aimed mainly at backpacker tourists. When I first visited Göreme in 1984, there were three small *pansiyons* or guest houses, plus one small restaurant. The second half of the 1980s decade saw the first major period of business proliferation, and by 1989, there were 50 *pansiyons* in the village, plus a few restaurants, tour agencies and carpet shops along the main street. Developing a reputation as a hub for international back-packers, Göreme featured in the main backpacker guidebooks such as *Lonely Planet, Rough Guide*, and *Guide du Routard*.

In 1990, the Göreme municipality undertook some major construction in the village centre, with the building of a bus station and taxi rank, plus an adjoining shopping and café complex. By the end of that decade, most of the previously dusty main roads running through Göreme had been tar-sealed, and some trees were planted, so the centre looked substantially different from the dusty village I had first visited a decade and a half earlier. By 1999, the number of *pansiyons* accommodating backpacker tourists had gone up to around 60, plus there were approximately 15 restaurants, 15 tour agencies, 5 or so bars or discos, 15 carpet shops and numerous other souvenir shops and stands, a couple of horse ranches running horse-riding tours, and one hot-air balloon operation. In *Living with Tourism (1)*, I wrote

> During the high season of the summer, the central area of Göreme village buzzes with tourist activity: tourists wander the streets and fill the many restaurants and shops that are there to service them; the little bus station in Göreme centre is crowded as hordes of tourists arrive off overnight buses… Somewhere between 1,500 and 2,000 international 'backpacker' tourists, equal to the number of local residents living in the village, stay in Göreme every week during the busiest summer season.
>
> (Tucker, 2003, p. 10)

Despite these tourism developments, interestingly, Göreme remained almost unrecognised as a tourism destination by Turkish officials – and academics – during the 1980s and 1990s. I remarked in *Living with Tourism (1)* that I was dismayed that tourism researchers I met in Istanbul and Ankara in the mid-to-late 1990s, along with officials I spoke to who worked in Cappadocia regional tourism offices, wondered why I was researching tourism in Göreme when, according to them, there was *no* tourism in Göreme. This perception may have been because most of Göreme's tourism businesses were too small to require registration, plus some of them were semi-informal businesses, and were thus not counted in national or regional statistics. As in many other countries around the world, backpacker tourism was not considered significant in terms of its contribution to national, nor indeed local, economies. Yet, it was precisely the particular characteristics of backpacker tourism, I argued in *Living with Tourism (1)*, which allowed Göreme's tourism to develop in a successful, or sustainable, manner. With the majority of businesses being small-scale and locally owned, and many of the pansiyons established in converted cave-houses, Göreme's tourist accommodation was particularly vernacular in character which, along with the other tourism services, aligned well with the backpacker tourists' liking for authentically local and unexpected experiences and interactions (Tucker, 2003).

Moreover, as I discussed at length in *Living with Tourism (1)*, the simple and informal character of their earlier tourism businesses placed the owner–entrepreneurs in a position of 'hosts' to their tourist 'guests', with *pansiyon* and restaurant owners being the main providers of what they considered 'Turkish hospitality', serving meals, taking tourists on trips in their cars, and singing Turkish folk songs when the tourists gathered in the evenings. Within this convivial relationship, the *Göremeli*[1] hosts were able to resist and divert some of the tourism representations placed upon them as 'traditional' or 'backward', and instead engaged in irony and playfulness, such as through Flintstones characterisation in their touristic performances, to negotiate a suitable identity for themselves and their place in the face of tourists and tourism. I wrote in *Living with Tourism (1)* (p. 75): 'Contrary to being "peasants without pride", as depicted by Schiffauer (1993) in his discussion of the effects of outward migration on Turkish village life and villager identity, the people of Göreme have retained, or even gained, a sense of pride'. Overall, tourism and the new opportunities to prosper it provided had significantly lessened any earlier imperatives to migrate out of the village, and so, as I wrote in the latter part of *Living with Tourism (1)* (p. 179), 'the village, while becoming divided into two separate realms, has in many ways stayed more intact and full of hope than it might otherwise have done'.

However, there was no doubt that significant tensions and clashes were also arising, both within the tourism business realm and in the daily lives of Göremeli people. These tensions were often borne out of a continuous sense that the tourists coming to Göreme were too few in number for everyone to earn a decent living from them, and I wrote about the increasingly aggressive and often violent competition among villagers in their clamouring to secure their share of the limited business available. I also wrote about there being an increased sense that 'hospitality, which lies at the centre of the villagers' identity, is gradually being abused and eroded' (p. 179). Nonetheless, I argued on the other hand that the idiom of the host–guest relationship was being successfully 'used to redress the inequitable potential for tourist behaviours and representations, providing villagers with a sense of control over the tourists in their village, while providing tourists with the interactive and serendipitous experiences they desire' (p. 182). Thus, from my in-depth examination of the dynamics between Göremeli hosts and their backpacker guests during the latter half of the 1990s, I concluded that tourism in Göreme, while being far from unproblematic, was likely set to continue as an example of successful, community-led, tourism development for some considerable time more.

In a brief postscript section at the end of *Living with Tourism (1)*, I referred to short visits to Göreme I made in 2000 and 2001, during the period that I was working on converting my PhD thesis into the manuscript for publication as *Living with Tourism (1)*. In that postscript, I wrote about observing: 'Hoards of backpackers wander(ing) the streets from early morning until the early hours of the next day, when they drunkenly make their way back from the Flintstones bar to the Flintstones, or some other, cave-pansiyon' (p. 187). I also wrote about the landscape becoming increasingly constructed *for* tourism, such as creation of a specified sunset-viewing location on a ridge overlooking the village and increased signposting erected by the National Park authorities to guide tourists while hiking in the valleys surrounding the village. I noted that the numbers of tourism businesses had increased, with a handful each of new pansiyons, restaurants, tour agencies, shops, and cybercafés, and I remarked that 'the mayor's "improvements" in the village centre, such as fancy street lamps and garden areas planted with flowers, were making Göreme appear less and less like the dusty village that it appeared to be even just a few years ago' (p. 187). I suggested that the developments that might most significantly influence the shape and possibilities of Göreme's tourism going forward were the opening of the new Kapadokya Airport, together with the burgeoning usage of the Internet in tourism, that is, usage both by businesses to advertise their services and by tourists to communicate with home. These developments appeared to align with an apparent global trend around the turn of the millennium towards more specialised forms of tourism, with an upsurge in independent tourists seeking vernacular and interactive experiences and services similar to the backpackers already visiting Göreme, but who might want somewhat more upmarket accommodation which they would book directly through the Internet. I wrote in the postscript that these new developments and trends would align well with what Göreme had to offer and that, therefore, the Göreme villagers would likely be able to prosper financially as well as retain a sense of place and control in their village. Overall, I concluded, 'the future for Göreme is looking pretty good, and a sense of hope is especially strong among the younger entrepreneurs' (Tucker, 2003, p. 188).

Longitudinal ethnography and my returns to the field

Very shortly after my September 2001 visit to Göreme referred to in the postscript of *Living with Tourism (1)*, the events of 9/11 took place. While the events had an enormous impact on tourism globally, with both heightened security and increased fear around flying, the Middle East region was especially affected in the aftermath. Since Turkey was associated with that region, at least in the eyes of Western tourists, Turkey's tourism industry experienced a dramatic downturn. I feared that Göreme and the wider Cappadocia region's tourism would be especially hard-hit, given its location significantly further east than most other tourism places in Turkey. Hence, when I was able to take half a year's research leave in 2005, I decided to spend much of it based in Göreme so that I could see what had been going on there in that interim period since I had conducted my last substantial fieldwork for my PhD and *Living with Tourism (1)* book. During that 2005 fieldwork, I bumped into an old friend Ali, an English and French-speaking tour guide. He was pleased that, even though I had moved to live in New Zealand by then, I had returned to Göreme to continue doing research there. He told me that he saw my return and continued interest in Göreme as "a kind of fidelity".

The idea of fidelity boosted my keenness to continue researching tourism and social change in Göreme as a long-term project. That interest was also spurred by my noticing that year, 2005, many new developments and changes occurring. There were many more village women visible in Göreme's tourism realm than there had been previously, particularly those working as cleaners and housekeepers in the increasingly upmarket pansiyon and hotel businesses. I hence set out to examine this particular point of change – by looking at what boundaries of propriety had necessarily shifted to allow women to work and be in tourism spaces. I became interested in *how* those boundaries might have shifted, what the role of tourism was in these changes, and also what

Figure 1.1 Göreme village.

the changes meant for gender relations more broadly (see Tucker, 2007 and Chapter 6 of this book for an in-depth discussion of these changes). Another notable change I observed during that 2005 fieldwork was the burgeoning hot-air ballooning industry in and around Göreme. During my earlier PhD fieldwork, in late 1997, I had spent some months living with the owners of the first – and only at that time – hot-air balloon company in Göreme. I therefore got up close to the goings-on regarding ballooning tourism in the sector's early days. Hence, the growth and changes that had occurred during the first half of the 2000s tweaked my interest and inspired me to continue to follow the twists and turns the sector would take over time (see Chapter 3 of this book).

While the fieldwork visit in 2005, after four years of absence, gave me the impression of a 'jump' with regard to certain changes, in the years that followed, I returned more frequently, going to Göreme every year or two for periods ranging from 1 week to 4 months. With this pattern, there tended to be fewer jumps and, rather, I developed a sense of continuities in my research whilst also being able to observe discontinuities or incremental changes as they were occurring. I thus was able to gather a sense of comparison, not 'between places, but... in the field itself, in a multitude of "ethnographic presents", encountered most meaningfully during regular returns to the field' (Hviding, 2012, p. 224). I was also able to see how particular events, or happenings, even those seemingly-at-the-time small and inconsequential, might trigger other processes of change which, through their accumulation, gradually shifted the boundaries of what was considered doable or "normal". As Guyer et al. (2007, p. 7) have described, the complexity of the relationship between time and social change is 'not necessarily hierarchical in that the larger (longer) processes shape the smaller or shorter term ones; and not necessarily neatly divisible without remainder'.

Repeated 'returns to the field' have become increasingly common amongst anthropologists (Howell and Talle, 2012), thus arguably changing the face of anthropology in that longitudinal research 'has held up both *change* and *persistence* to be regular features of human society' (Royce and Kemper, 2002, p. xvi). Thomson and Mcleod (2015) go so far as to suggest that a 'temporal turn' has recently taken place within the social sciences, allowing 'for an understanding of social phenomena in greater time perspective' (p. 244), although arguably many anthropologists have been doing longitudinal studies for decades (for example, Elizabeth Coleson's significant longitudinal work in Zambia). Some anthropologists interested in tourists and tourism, too, have produced ethnographies about tourism in one place over a period of time. For example, Stroma Cole's *Tourism, Culture and Development: Hopes, Dreams and Realities in East Indonesia* (2008) is a study of tourism development and social change in Flores, Indonesia over 15 years. Beth Notar studied Dali in China over ten years for her book *Displacing Desire: Travel and Popular Culture in China* (2006). Kathleen Adam's *Art as Politics: Re-crafting Identities,*

Tourism, and Power in Tana Toraja, Indonesia (2006) was researched over two decades, and Kenneth Little, too, collected stories of 'make-Belize' over several years for his book *On the Nervous Edge of an Impossible Paradise* (2020). Among these, Notar's (2006) *Displacing Desire* and Cole's *Tourism, Culture and Development* (2008), are most similar to my earlier Göreme work, since they, like *Living with Tourism (1)*, are studies of the transformations which ensued as places became backpacker destinations during the 1980s and 1990s; they too were studies of what occurred in places that "lonely planeteers", as Notar called them, were drawn to.

Indeed, I feel very fortunate that I have been able not only to experience, observe, and write about Göreme's tourism from its very early days – as it became a backpacker tourism place, but also right through to its becoming a global *Instagram* sensation and featured in *Tripadvisor's* 'Top 10' list of places to experience in 2019. I have been able to observe what might be considered Göreme's 'tourism destination evolution' (Brouder et al., 2017) or, according to Butler's (1980) classic theory, its 'tourism area life-cycle', although I am not so sure that either of the terms – 'evolution' or 'life-cycle' – are apposite in relation to the story/stories of tourism and change in Göreme. My observations have taught me that tourism development and its associated social change cannot be explained as occurring in some sort of linear 'unified progress-time' (Tsing, 2015, p. 34). To the contrary, since change occurs as a tension between global and local 'forces' and fluxes, it is often paradoxical and thus manifests as deep contradiction and ambivalence in peoples' experiences of everyday life. Nor can tourism development and social change be portrayed in a 'before and after' manner, for example, before-tourism and after-tourism or, as is common recently, depicting a place as switching from a state of 'undertourism' to 'overtourism' – 'where hosts or guests, locals or visitors, feel that there are too many visitors and that the quality of life in the area or the quality of the experience have deteriorated unacceptably' (Goodwin, 2017, p. 1). Indeed, while the term overtourism may appear apt in relation to Göreme these days, it is rightly argued by Jóhannesson and Lund (2019) that describing a place as being in a state of 'overtourism' fails 'to fully grasp the mobility of place and multiple entanglements through which tourism emerges and affects society and culture' (p. 92). This is precisely why, as Anna Tsing (2015) suggests, it is so important to 'listen to and tell a rush of stories' (p. 37). Tsing reminds us that, since there is little point in attempting a 'summing up' or – even more impossible – attempting 'to compute costs and benefits… to any "one" involved' (p. 34), we might instead 'revitalize arts of noticing' (p. 37) and embark on listening to, attending to, the 'cacophony of troubled stories' (p. 34) as the best way of developing our knowledge practices in relation to the contemporary neoliberal world.

Thus, through accounts of change-happenings in Göreme during the past two decades, my aim in writing this second book is to attend to how the intensities of this increasingly globalised and apparently 'accelerated world' (Eriksen, 2016) are experienced and negotiated as they manifest in practices of the everyday. During my many 'returns to the field', frustrating as it is to come in and out of the field, in order to experience how Göreme's *Living with Tourism* continues to play out, when I am there, I jump into life as much as I possibly can. I spend time with my Göremeli surrogate families in their homes and gardens; I visit old friends in their businesses; I walk around and bump into old plus more recently made acquaintances. As well as catching up on past happenings and events that I had missed while away from Göreme, I catch up on the present, seeing what people are up to, what their concerns are, what they think about their life, about Göreme, and the world. In this way, I gain an understanding of how people go about negotiating the undulations that life throws at them, and which in turn bring about new openings for possibility that were not previously available nor anticipated. While I do perform a tracking or monitoring of sorts in regard to changes occurring over time, during fieldwork visits, I like to get into the rhythms of Göreme life; the rhythms of each day, and of the week, as well as the rhythms of the seasons, which are strongly felt both in relation to tourism and in the experiences and practices of Göreme daily life.

My favourite time to be in Göreme is in September and October, when the summer is turning into autumn. This is the time when villagers who still tend their fields – or gardens as the more common term used – are busy harvesting the produce and working in all the different ways in which they preserve that produce in preparation for winter. I help them out at their gardens harvesting apples, grapes, and walnuts, and then join in the work as families and neighbours gather around small fires they have lit to boil *pekmez* and other sticky preserves.[2] Up until a decade or so ago, at that time of year, the neighbourhood streets would become places of food-work (see Chapters 5 and 7), with groups of women using the communal ovens dotted around the old neighbourhoods to bake stuffed breads, pastries, or to roast pumpkin seeds, and *pekmez*-fires would be everywhere. While these practices are no longer common due to the increased bustle of tourist rental cars, delivery vans, and airport shuttle buses in those neighbourhoods, even in recent years, I might still come across a group of women cooking *pekmez* or *acılı* (a spicy pepper sauce) over an open fire tucked away in a small side street, and when I do, I stop for a while to help out and so as to be able to observe and chat about daily life in the neighbourhood and Göreme.

If I visit during the hot summer months, I frequently get up and out early, either to help my friends and neighbours do whatever work their gardens need before the heat of the midday sun, or I go somewhere to watch the hot-air balloon happenings which, during the summer months, occur as the sun is rising at around 5 a.m. During the hot midday hours in summer, Göreme's streets tend to be fairly quiet; people are in their houses or hotels, either having gone back to bed for a couple of hours after gardening or ballooning, or just resting in the cool of a cave-room or the shade of a courtyard wall. Come late afternoon, the temperature has usually cooled off enough for tourists to go out and peruse the shops, maybe have a beer, or make their way slowly up to Aydın Kırağı, a ridge overlooking Göreme referred to as 'Sunset Viewpoint' – or often, these days, 'Sunrise Viewpoint' – before eventually finding a restaurant to have dinner in.

Figure 1.2 Tourists at Sunset Viewpoint as the sun sets over Göreme.

I enjoy the late afternoons for walking around and seeing whom I will bump into, or sometimes I sit and buy a Turkish coffee on the main street so that I can watch the world go by; Russian tourists in mini-skirts and tight low-cut T-shirts, a Chinese couple in wedding regalia having just finished a photo-shoot for their social media postings, an elderly neighbour – dressed in *yemeni* (headscarf) and *şalvar* (baggy trousers) – waving to me as she passes on the back of her husband's moped, the council fly-spray truck slowly driving around the streets spraying toxic fumes behind it as it goes in an attempt to dampen down the latest outbreak of flying insects. If I choose to keep away from the busy main street and instead go for a walk up and around the residential streets of the old neighbourhoods, until quite recently I would encounter groups of women who, in that cooler part of the day, had come out to sit on doorsteps or cushions placed in the street outside their houses, sometimes with a pot of tea to share and likely some gossip too (see Chapter 5 of this book for an account of changes to neighbourhood life and neighbouring practices). I would often join such groups, knowing many of the women anyway, or even if I did not know them, I would invariably walk up and say a few words of pleasantry and, upon realising I speak reasonable Turkish, they would pour me a glass of tea and invite me to sit down. This way, over the years, I have got to know more and more people every time I have visited Göreme, and sitting with people in this way has allowed me to get a sense not only of their daily lives but also of how they participate in and experience the changing world around them and of which they are a part. When I do these walks-around, or sittings-around, I always carry a field-note booklet with me, as well as a small English–Turkish dictionary so that I can look up unknown vocabulary during conversations. I scribble notes whenever I get a chance, and when I explain to new acquaintances the purpose of my notes, they invariably share further reflections on Göreme's tourism and change, as well as offering to take part in a later interview or suggesting they introduce me to someone else who would be relevant for me to talk to.

The rhythms of daily life in Göreme are felt strongly also in relation to the cyclic passing of each week. Monday has always been the day to get jobs done, going to the *Belediye* (Municipality Office/Town Council) office to pay bills, or driving or taking a bus for the approximately ten kilometres into the regional town of Nevşehir to visit the bank, get things fixed, or visit the weekly market there. Wednesday is market day in Göreme, when several market vendors from elsewhere come and set up stalls. Whilst not a big market, there are usually several fruit and vegetable stalls, a watermelon truck, cheese, olives, honey and nuts, as well as stalls selling clothes, toys, hardware, and kitchenware. Attending the Wednesday market is inevitably a sociable occasion, as it is where I bump into Göreme women I know who like to advise me on which tomatoes or aubergines to buy, and I often meet hotel or restaurant-owner friends as they do last-minute shopping for their menus. Friday is the next day when a few vegetable and other produce vendors come to Göreme, but on this day, instead of setting up in the market-place adjacent to the 'new neighbourhood' where most of Göreme's inhabitants reside these days, they set up across the road from the large central mosque which becomes a hive of activity around Friday prayer-time. Nowadays, particularly in the summer months, with Turkish and Middle East tourists as well as incoming workers adding to the local population of worshippers, there are so many people attending the main mosque for Friday prayers that latecomers have to lay down their prayer mat outside the mosque.

On Fridays, I often make sure to time a visit to the fruit and vegetable stalls so that I can enjoy the almost festive atmosphere as the men pour out of the mosque. Due to gendered spatiality in Göreme society, it is inappropriate for me to go into the mosque or the teahouse, so being around the central mosque area as it empties, browsing the fruit stalls or sitting on a park bench eating an ice-cream, is an opportunity to catch up with many of the tourism businessmen whom I know as both acquaintances and friends. The conversation with Mustafa referred to at the start of this chapter occurred during one such meeting. Although usually brief at the time, these meetings often lead to something else: an invitation to go and view someone's latest hotel renovations; a trip to someone's field to help pick tomatoes and water the almond trees; or a chance to confirm an arrangement to conduct a more focused 'interview' chat about tourism. On occasion, these encounters wind up with me being invited to join some of those friends at the teahouse. As it is men's space, I tend not to feel comfortable there, so I only stay long enough to drink just one glass of tea. In recent years, I have found that this usually is long enough for the conversation to turn to nostalgic reminiscing among the men I am sitting with. At such times, I too may find myself indulging in nostalgia.

In recent years, the central streets tend to be busier and noisier on Friday and Saturday evenings than on mid-week days. Several large restaurants, which cater to the increasing number of *yerli* (local/domestic) tourists who come to Göreme especially at weekends, compete with each other in the volume of the live Turkish music they provide. Twenty years ago, there was no noticeable difference in tourist numbers between mid-week and weekend days, since international backpackers would come into town, and leave, on any day of the week. With the significant rise, especially during the past decade, of domestic tourism in the Cappadocia region, along with a surge in weekend day-trippers from nearby cities and towns, the volume of car and foot traffic around Göreme is noticeably elevated during weekends. As well as the central streets, certain vantage points, and scenic areas for balloon-watching or sunset-viewing tend to be more crowded during the weekends, especially during the summer months and festival, or *bayram*, periods. When the summer or festive holidays are over, the week days and weekends are most strongly differentiated by Göreme's primary and intermediate schools being open or closed. During the week, the school bell rings out and uniformed children walk between home and school carrying backpacks full of books. This is a reminder that while Göreme has become a prominent tourist destination, it is also a place of home for *Göremeli*, and others increasingly, to live their everyday lives.

All of these rhythms of daily life – days, weeks, years – have changed considerably over the past 20 years, both in terms of how they are practiced, or lived, and in terms of affect, or the way that being in Göreme *feels*. I find it mildly amusing that in the postscript of *Living with Tourism (1)*, I wrote that during my visit in 2001, I had sat with my friend Abbas on the front steps of his tour agency office and, as the main street seemed so busy with all the tour buses trundling past, Abbas remarked "Istanbul gibi oluyor" (it is becoming like Istanbul); back in 2001, little did people know how things would continue to change over the course of the next two decades. As well as the changing feel of the central parts of Göreme, the residential neighbourhoods have seen significant change during the past two decades, as I discuss in-depth in Chapter 5. Previously considered the *kapalı*[3] (enclosed) parts of the village and therefore the place for women – also *kapalı* – to comfortably live their daily life, the old neighbourhoods have been transformed into places of tourism during the past two decades, becoming filled with hotels, restaurants, rooftop bars, and *Instagram*-photography terraces.

Making repeated fieldwork visits to Göreme over several decades, this research is inevitably a product of the relationships I have built with people there over the years. In many of those relationships, my initial primary positions of researcher and tourist have transitioned to that of long-term friend, or surrogate family in some cases: I have become a "sister", "daughter", "niece", or "sister-in-law" to several households or individuals. My relationship/s in and with Göreme have also inevitably re-formed as others' circumstances, and my own, have changed. When I was doing my postgraduate studies and fieldwork for *Living with Tourism (1)* during the 1990s, people knew me – and looked after me – both as a student and as a single woman from Scotland. When I returned to Göreme in 2005 for the first major period of fieldwork for this current book, I went with my New Zealand husband and our baby son. We borrowed an empty cave-house to stay in and were lent a crib for our son to sleep in, along with other furniture, so that we could set up "home" in the neighbourhood of Aydınlı Mahallesi (see Chapter 5). We made similar living arrangements during subsequent longer fieldwork visits, and when I made shorter visits in-between those lengthier bouts of fieldwork, I either was a guest in friends' houses or I would be given a room in one friend's or another's hotel. As time went on, hotels became full much of the time and the pressure on accommodation more generally made it increasingly difficult to borrow or rent houses or rooms for longer stays. In the early 2010s, I had the opportunity to buy a cave-house in the lower part of Aydınlı Mahallesi. While providing me with the surety of having somewhere to stay during visits, owning property in Göreme has inevitably created a further change in the way I relate to and live in Göreme when I am there; for example, I have become acquainted with the local carpenter, the gas-man, and the owner of the hardware stall at the weekly market. Since I am not in Göreme for much of the time, also, I let the lower part of the house to the owner of a carpet shop business who was among the first to have the idea of opening a shop in the old cave-house neighbourhood. I thus played some part in the start of a new trend to open tourist shops in the narrow winding streets of the old neighbourhoods.

In line with reflexivity in feminist anthropology scholarship, I endeavour to explicitly place myself as the researcher and author of this book. Inevitably, not only has Göreme changed over the years, but so have I in the way I work, research, and write. Ethnographic research in tourism studies was fairly novel when I undertook the work for *Living with Tourism (1)*, and it tended to engender a rather tiresome, and, in my view, senseless, debate regarding the merits and demerits of qualitative versus quantitative research. It was around that time that an ontological turn occurred in tourism studies, a turn away from the rather 'stale, tired, repetitive and lifeless' (Franklin and Crang, 2001, p. 5) writing about tourism. I embraced this turn in subsequent years and, wishing to contribute towards an altogether more 'exciting body of work that speaks to the multi-layered phenomena of contemporary tourism spaces, mobilities and encounters' (Everingham, Obrador and Tucker, 2021, p. 71), I began to write more reflexively, particularly in relation to matters of emotion and affect (shame, empathy, hope, and fear) in tourism encounters (for example, Tucker, 2007, 2009a, 2011, 2014, 2016a). I also based some of my writings overtly around particular tourism encounters in which I had been involved, attending to encounters as performative sites of affective negotiation. I found that doing so entailed something of a re-learning of the art of noticing, or of paying attention (Savransky and Stengers, 2018), during my many return visits to Göreme.

Along with telling stories of individual people, and, in some cases, families, as well as quoting extracts from the more purposeful interviews I have conducted during multiple bouts of fieldwork, I will frequently focus on stories or anecdotes of encounters, recorded in over 20 field-note diaries I have filled over the years. In this book, I use these jotted-down encounters, conversations, and observed 'happenings' more in some chapters than in others, often as the jumping-off point for a discussion, or to illustrate an argument I am making, and always, also, thereby hoping to present myself overtly as the ethnographic lens. The more purposeful interview conversations were sometimes conducted in English, such as with men who work in tourism and whose English level surpasses my Turkish, while interviews with Göreme women were generally conducted in Turkish and sometimes with the help of a bilingual research assistant to provide surety of comprehension of nuanced meaning. While these interviews formed a wealth of rich material, as did the recorded observations, encounters, and a great many casual conversations I have had during times spent in Göreme, I acknowledge that both my knowledge and my writing cannot in any way convey the entirety of experiences and intensities which living with tourism in Göreme comprises. As Donna Haraway (2003, p. 31) has put it, 'the knowing self is partial in all its guises, never finished, whole, simply there and original'. So, too, is my longitudinal research and my storying of tourism and change in Göreme.

Unlimited change

Towards the end of *Living With Tourism (1)* I summarised what I saw as both continuity and change engendered in Göreme during the first two decades of tourism there. As alluded to above, there was an overarching sense that, due to the relative economic stability households were able to attain from tourism, Göreme had likely 'stayed more intact and full of hope than it might otherwise have done' (Tucker, 2003, p. 179). Whilst this 'intactness' suggested continuity and longevity, the vibrant 'cultural negotiations between tourists, villagers and the various authorities' (p. 166) at play simultaneously professed the inevitability of change: 'marriages; deaths; a sense of the transition from a poor past to a hopefully prosperous future; changes in the season and the yearly opening, closing, reorganisation and changing in appearance of the many tourism businesses' (p. 166). However, with all such changes concurrently juxtaposed with continuity, I referred to what I saw as the *limits* of change, whereby the different realms of living with tourism in Göreme seemed always to tug on each other so that no aspect of change could push itself outwards too far. For instance, while 'the limits of the codes of honour and shame concerning local women are stretched when women attend a wedding party by the pool of the Turist Hotel', 'men's fun and sexual relations with tourists are checked and inhibited by their moral ties with their families and their "home" lives' (p. 182). Hence, although I saw at that time the limits, or edges, of change stretching and being stretched, the boundaries of moral propriety, along with the ingrained moral economy, appeared to keep a hold on things so that the metaphorical frontiers of change would shift only in small, incremental steps. Moreover, the change occurring as Göreme became a place of living with tourism in the 1980s and 1990s was generally hopeful change (see Chapter 7 in this book for discussion on hopes). By and large, with tourism offering possibility and opportunity, there was longing for such change.

In conjunction with this longing for change, however, there prevailed a view of the opportunity that tourism might provide as limited. In other words, there was a continuous sense of there being too few tourists, out there in the world, who would want to come to Göreme. This view was enhanced by periodic downturns in tourism caused by events or happenings outside of Göreme, and often outside Turkey, such as the Persian Gulf War of 1990–91 and the '9/11' attacks on the World Trade Centre and Pentagon in 2001. Indeed, prior to Covid-19, tourism businesses and livelihoods in Göreme experienced several significant downturns, usually caused by political events occurring at the international, national, or regional level. While the Covid-19-related tourism shock provided a worldwide reminder that local and global processes and happenings are always so intricately intertwined that we could never properly refer to one without also talking about the other, for Göreme's tourism, Covid-19 was simply another downturn of many. As with all tourism at the local level, Göreme's tourism is always caught up in, and caught up *with*, not just major global or regional events, but the ebbs and flows of broader economic conditions and societal fluxes, as well as tourism trends. Particular global trends which have greatly affected Göreme's tourism during the past two decades include the emergence of mobile digital technologies and social media, and changing tourism markets, such as the rapid growth of outbound tourism from China. Shifts in tourism-related behaviour and meaning such as tourists' use of social media to project their 'perfect self(ie)' have also had a significant influence on the ways in which tourism in Göreme has mutated and developed, as I will discuss in-depth in Chapter 4.

It should not be assumed, however, that these global influences are altogether weightier than local processes and practices. While earlier analyses of change dynamics in tourism destinations largely focused on the role played by tourism demand and the changing preferences of tourists, there is more recent acknowledgement that it is not possible to study destination change 'without also including social, cultural, economic, and environmental changes and challenges' (Brouder et al., 2017, p. 8). Indeed, tourism studies thinking has come to view the global not as 'a separate force impinging upon "the local", but instead part and parcel of everyday life' (Leite and Graburn, 2009, p. 53). In line with this view, Milne and Ateljevic (2001, p. 379) suggested that the business of tourism always necessarily incorporates 'both exogenous forces and the endogenous powers of local residents and entrepreneurs'. A particularly important aspect of such 'endogenous powers' in Göreme's case are the forms of tourism business practice and sociality engendered by what George Foster (1965) identified as an 'image of limited good' prevalent in the psyche of peasant societies. This 'image of limited good' manifested in Göreme's *Living with Tourism* in a variety of ways so that, as I will explain in-depth in Chapter 2, tourism enterprise became embedded within – and thereby maintained – many aspects of peasant moral economy and sociality, rather than overriding them.

While major global events, such as the Gulf War in 1990–91 and, ten years later, the events of '9/11/2001' served to confirm 'tourism as limited good' beliefs, the decade or so which followed '9/11' developed as a period of relative stability. By 2010, Göreme – along with the broader Cappadocia region – had become the hot-air ballooning centre of the world, and there seemed to be no shortage of tourists keen to visit. There was a sense that Göreme was in its tourism heyday; business was booming, property was being sold and bought as if Göreme was the board in a giant game of *Monopoly*, and many Göremeli families had become very wealthy indeed. This tourism growth continued into the following decade, with the combining of social media and tourism further accelerating the growth already occurring spurred by the burgeoning hot-air ballooning sector. While that following decade (2010–19) threw up several curves and bumps for Turkey's tourism, including terror bombings in Istanbul, the Syrian crisis prompting millions of refugees to enter Turkey, and the 2016 attempted coup, Göreme's tourism seemed each time to bounce back bigger and stronger. As the decade went on, the earlier notion of tourists and the potential 'good' from tourism as limited was gradually undone; now tourists and the potential wealth they might bring appeared not only to be ever extending, but they also came to be seen as extend*able*.

The shift, whereby the horizons of the good are expanded and expandable, inevitably and fundamentally changed social relations, and this change has accelerated particularly since the turn of the millennium. While the spectre of neoliberalism had already loomed as far back as the early 1980s, its presence was not so apparent then – largely due to Göreme's 'protection' within the Göreme National Park, as mentioned earlier. While the Turkish government's 1982 Tourism Encouragement Act prompted market liberalisation processes in the tourism developments in other Cappadocia towns outside the perimeter of the national park, the limitations imposed on building and business size within the park meant that, in Göreme, an image of limited good continued to underpin the peasant-cum-tourism moral economy and sociality as tourism enterprise was developed there. Yet, during the past two decades, while remnants of an image of limited good and some of the aspects of peasant sociality it engenders remained, a significant shift occurred whereby tourism started to be seen as promising *unlimited* 'good'. I see this shift as being at the core of Göreme's story over the past two decades.

Indeed, June Nash (2007) has suggested that much of the world's peasant populations 'are now facing elimination in the global markets of the third millennium' (p. 4). According to Nash, this is due not only to the spread of global neoliberal market economics per se, but to 'the spectre of the unlimited good' associated with that spread. Nash argues that an image of limited good 'balances the collective needs of a population with available land and resources, while the opposing spectre of unlimited growth cultivated by neoliberal market economics benefits only a few' (2007, p. 4). With Öztürk et al. (2018, p. 248) similarly suggesting that peasant sociality is 'based on the struggle to balance (rather than exploit) human–nature relations', it becomes apparent that an image of limited good and the associated peasant moral economy which engenders a sociality of balance and reciprocity resonate a particular poignancy in today's (somewhat troubled) world. Moreover, the peasant sociality of limited good necessarily acknowledges that there is no such thing as self-containment of the individual. This is in sharp contrast to the assumption of self-containment argued by Anna Tsing (2015) to be 'the hallmark of modern knowledge' (p. 33), and which 'allows us to fantasize – counterfactually – that we each survive alone' (p. 29). As Öztürk et al. (2018) explain in reference to the (re)production of a contemporary, or new, peasantry in Turkey, 'the peasantry actively commits to reproducing itself, its living arrangements, neighbourhood relations and communal ties' (p. 251).

It is pertinent, then, to consider how the image of limited good and its associated sociality become eroded, and how a neoliberal capitalist spectre of unlimited good comes to manifest, in this case through tourism. Indeed, how does the promise (spectre) of 'unlimited good' become so irresistible, even if it might thoroughly blow apart the 'living arrangements, neighbourhood relations and communal ties' that have long been actively reproduced (ibid.)? It is the neoliberal notion of unlimited good that underpins the *spectre of unlimited change* I proffer as the title of this book. As I will go onto describe in the chapters that follow, not only have the tourism growth and developments during this period caused literal displacement of Göreme's residents, but they have resulted in what Cocola-Gant (2018) terms 'place-based displacement', whereby such a deep mutation of place has occurred as to cause even those residents who remain to experience *loss of place* (see Chapter 5). The emotive notion of solastalgia also comes to mind (see Chapters 7 and 8), explained by Albrecht (2005, p. 45) as the 'pain experienced when there is recognition that the place where one resides and that one loves is under immediate assault' (p. 45); this would appear to resonate strongly with Mustafa's expression of sadness outlined at the start of this chapter. This book tells the story/stories of how, for many, hopes that were previously enabled by tourism in Göreme have, during the past two decades, become a foreboding spectre of dreams come true. Overall, the book speaks not only to what happens when tourism in a place loses the sense of its own limits, but also to the paradox of continuing to drive the metaphorical car even when its brakes are showing signs of failing and, hence, when the spectre of unlimited change is looming.

Chapter guide

The book's chapters may either be read in order – from Chapter 1 to 8, or readers might choose to hop around individual chapters, depending on which chapter's focus grabs their interest in any particular moment. Most of the chapters engage a decadal time frame by storying the subject of their focus – balloons, tourists, neighbourhood displacements, changing women's lives – from the 1980s and 1990s to the 2000s and 2010s. Such a decadal approach, according to Guyer et al. (2017), is apposite to longitudinal research in that, 'above all, periods of decades correspond to the human world' (p. 6) as periods over which change dynamics play out. Rather than suggesting any sort of linearity of time or an ability to see discrete patterns of change with identifiable causations and effects, however, the different chapters, when put together – either side-by-side or one on top of another, should 'say that life is not just one story but a host of different stories, …[thereby] asserting the possibility that these stories can run alongside one another' (Ingold, 2011, p. 142). In other words, my intention is that, whether read in order or hopped around back and forth, the chapters and their stories *together* will depict the layering of multiple interwoven trajectories always simultaneously at play. As Jóhannesson, Ren and van der Duim (2016, p. 78) rightly suggest in relation to tourism entanglements, 'ultimately, everything is in everything else'.

Chapter 2, "Four decades of *Living with Tourism*: from limited to unlimited good", tells of tourism developments and entrepreneurship dynamics in Göreme since their humble 1980s beginnings to the year 2019, when growing opulence and wealth were coupled with the increased inequality and displacement of local residents. In order to depict the interconnectedness of global fields of influence within the local and even personal levels, early in the chapter, I present portraits of six people, including some key characters who had started the initial tourism businesses – Göreme's first restaurant and tour agency – during the 1980s. Focusing in on Göreme's entrepreneurs being 'peasants-in-transition', the chapter discusses how tourism business at first became embedded within the peasant moral economy of 'limited good'. The chapter then goes on to discuss the 2000s and 2010s when, propelled by the burgeoning hot-air ballooning sector and increasing social-media-influenced fame, there was a shift to seeing tourism as promising 'unlimited good'.

In Chapter 3, "The ups and downs of hot-air balloons over two decades", I depict the balloons as principal protagonists, or 'movers and shakers', in Göreme's story over the past two decades. Following the hot-air ballooning developments over a period spanning 20 years, I give the chapter an explicit decadal structure, discussing various entanglements of the hot-air balloons across three main phases – circa 1999, 2009, and 2019. Within the three time-periods, I discuss various balloon characters, or configurations – 'Elite balloons', 'Money-maker balloons', '(Un)regulated balloons', 'Nuisance balloons', and 'Instagram balloons' – in relation to their tourism-related orderings and enactments. Following the emergence and decline of the different balloon characters allows me to consider the ups and downs of various balloon entanglements which ultimately led to the hot-air balloons coming, within two decades, to dominate the tourism landscape of Göreme.

Chapter 4, "In search of the perfect self(ie): making stories, narrating self", I focus on the tourists visiting Göreme and consider the changes, over different decadal periods, in what makes, and what affords, a good tourist story. The equivalent chapter in *Living with Tourism (1)* was entitled 'The tourists: in search of serendipity', and that chapter described how the predominantly Western backpacker tourists who stayed in Göreme during the 1980s–90s sought adventurous and authentic encounters and experiences. Since then, a gradual decrease in backpacker visitors coupled with an upsurge in both Turkish visitors and major new international markets has substantially changed the tourist profile in Göreme, along with the kinds of businesses and attractions created to service them. In addition to bringing new preferences and meanings in terms of what makes a *good* tourist story, new technologies – particularly in the form of mobile devices and social media platforms – have, during the past two decades, afforded new ways for these contemporary tourists to convey their stories and thereby to project their version of their perfect self(ie). I discuss in this chapter how these new tourists are viewed by local entrepreneurs and residents, concluding that with tourists' focus switching from the places and people they are visiting to them*selves*, the tourists too have become a key aspect of the spectre of unlimited change.

In Chapter 5, "Changing life in the mahalle: neighbourhoods and displacement", I discuss the proliferation of hotel construction and cave-house-to-hotel conversion in Göreme's old neighbourhoods (*mahalle*) which has led to the actual displacement of residents as well as to an overall sense of place-based displacement in Göreme. Beginning by discussing the gendered neighbouring practices which previously were at the centre of women's daily lives in Göreme, I then go on to explain how, in the rapidly gentrifying old neighbourhoods, the now-extensive conversion of residential houses into tourist accommodation businesses has prompted the departure of increasing numbers of residents, or neighbours, from these neighbourhoods. Furthermore, not only has there been an increased presence of tourist accommodation businesses in these neighbourhoods, but an influx of workers, vehicles, and the general bustle of tourists and tourism has manifested as a total loss of the familiar in Göreme, experienced as a sense of deep mutation of place which particularly affects women's daily lives.

Chapter 6, "Frontier of change: from being *Kapalı* to crafting new selves", focuses on the shifting gender roles and relations which have occurred during the past two decades. I explained in *Living with Tourism (1)* that, as tourism had developed during the 1980s and 1990s, Göreme and the wider Cappadocia region had remained a pocket of Islamic social conservatism. With codes of honour and shame upholding a gendered spatial separation so that women's lives and practices were spatially centred in and around the household, the tourism business initially became almost entirely the domain of men, and women were largely separated from tourists and tourism activity. However, during my returns to Göreme over the following two decades, I observed how local women became increasingly involved in tourism; many gaining employment in hotels and guest houses and some women embarking on entrepreneurial activities, such as making and selling jewellery to tourists, running cafés, and restaurants, or offering Turkish cooking classes. This chapter discusses the part that tourism has played, along with other aspects of social change and reforms regarding gender equality and education, in shifting the gendered spatial and moral boundaries in Göreme. The discussion outlines women's ability to 'craft new selves', either directly through tourism work and entrepreneurship, or indirectly through the increased wealth from tourism affording girls and young women improved education, career, and livelihood prospects when compared with previous generations.

In Chapter 7, "Sticky memories, hopes, and dreams," I form a discussion around reflection and affect to highlight the ambiguities and contradictions surrounding tourism and change. I use anecdotes of making *pekmez* – a sticky grape molasses – to bring into view a generational perspective on the potential loss, and the potential revitalisation through tourism, of particular "sticky" cultural practices. I consider how the urging of certain cultural practices to be revitalised and sustained, both for and through tourism, is juxtaposed against hopes of becoming something other; these hopes being highly gendered and varying between different generations. While the heritagised nostalgia within tourism – which acts to tie women's daily lives, bodies, and activities to a logic of sameness – does not go without some level of resistance, it can render somewhat invisible particular moral imperatives which have long acted to shape women's lives. Discussion of these conundrums and ambivalences highlights how sticky memories, hopes, and dreams work constantly to undermine and destabilise each other. While the younger people of Göreme grasp onto tourism's potential to improve livelihoods and enable their hopes of becoming something other, older generations tend towards nostalgia, so that even for those who have become very wealthy from tourism, the extreme change in Göreme manifests as a spectre of dreams come true.

In the book's final chapter, "Enough! (*Yeter!*)", I draw together threads from all four decades of Göreme's tourism and social change to develop key themes arising from the stories told in the chapters of this book. These themes are developed as three main crises – a crisis of enough, a crisis of hospitality, and a crisis of "our town". These three crises, together, manifest as a deep ontological crisis which looms over Göreme both as an ambivalent site of affective negotiation and as a spectre of unlimited change.

Whilst, as already mentioned, the book's chapters do not necessarily need to be read in order from 1 to 8, certain anecdotes, people portraits, and stories from earlier chapters are built upon, or referred back to, in later chapters. I use these people stories and anecdotes of encounters to give both a sense of life lived and of how the many different layers and trajectories of change are refracted in people's daily lives. Additionally, many of the anecdotes overtly include me, thus serving as a reminder that I am always situated in relation to the foci of my interest during fieldwork in Göreme, as well as in relation to my writing of this book. Of course, my portraits of people and their stories told in the different chapters are always partial. As already mentioned above, people and stories are always *more than* what I know or choose to write about them; it is inevitably an enormous responsibility to attempt to write others' stories at all. Where I have written about individuals, some names have been changed to protect privacy while others – where expressed permission has been given – are people's actual names. Some of the people written into this book also featured in *Living with Tourism (1)*. Proud to be protagonists in Göreme's *Living with Tourism* stories, among those people to whom I gave a pseudonym in the last book, many have requested that their real name be used in this second book. One such person, Osman, when discussing with me the matter of names used in the books, said: "We trust you Hazel; other people might write Göreme's story from their head, whereas you write from your heart". I hope that in what follows I do Osman's words justice.

Notes

1 The term *Göremeli* refers to a person being of or from Göreme, and is the status afforded a person either through birth within the village or through male lineage connected with Göreme. Women from elsewhere who marry into a Göremeli family also become Göremeli.

2 *Pekmez* is a sweet molasses traditionally made by boiling grape juice.

3 As will become apparent at various points in this book, the dual notions of *kapalı* (meaning covered or enclosed, and hence clean) and *açık* (meaning open but also unclean) are prominent and important terms in Central Anatolian village life. While strongly associated with gender relations and propriety (see Chapter 6), the terms also have spatial associations akin to a private/public duality. As Carol Delaney (1991) noted in her discussion of gender and cosmology in Turkish village society, villagers consider their village to be *kapalı* in contrast to the *açık* city and, within the village, the residential neighbourhoods are *kapalı* relative to the *açık* central and public parts of the village.

2 Four decades of *Living with Tourism*

From limited to unlimited good

> ...the notion of the unlimited good, while raising the levels of expectation for a better life, may also cultivate behavior that raises a specter [*sic*] of doom for many.
>
> (Nash, 2007, p. 41)

During a visit to Göreme in September 2010, while chatting with Abbas about the extent and rapidity of tourism growth in Göreme over what was then three or so decades, he referred to his own and his friends' role in starting tourism by saying: "We cut a new path and now it is a huge highway"! Abbas had already opted out of tourism by then; having sold his tour agency and bought a tractor, he had decided to return to farming. The produce from his gardens would maintain his family's subsistence while the modest pension he received from the government along with the rent from letting the ground floor of his house would allow him to pay bills and taxes, buy household goods when needed, and – most importantly – see his son through school. Following that conversation, every now and then during the next decade (2010–19), Abbas would again make comment about how, although he had been a pioneer by opening the first tourist restaurant in Göreme, it was others – some of them his good friends – who had become wealthy from tourism. Meanwhile, when I chatted with those friends of Abbas who continued to strive to increase their wealth even further from the tourism growth happening around them, they would often tease Abbas with a friendly mocking banter about how he had returned to his gardens and become, once again, a 'çiftçi' (farmer). Sometimes, when having these conversations, I would test connotations and meanings by asking whether Abbas had reverted to being a '*köylü*' (meaning 'peasant' and also translating as 'villager'). He and his friends would generally reply that, no, he was a farmer now, not a peasant: "In any case", they would say, "how can he be *köylü* when Göreme is no longer a *köy* (village)"?

The category of 'peasant' is invariably considered to centre on smallholder and subsistence farming. During my fieldwork in the late 1990s, many people, including Abbas, were still growing their own wheat to make *bulgur* and

DOI: 10.4324/9781003011200-2

flour, pasta, and bread. As well as providing for self-provisioning throughout the summer and autumn months, produce from the villagers' gardens was preserved in various ways for winter consumption. Most village households produced grape molasses (*pekmez*), vinegar for pickling vegetables, fruit preserves, and dried flatbreads. The larger agricultural and food-production jobs were undertaken in *topluluk* (neighbourhood groups); forms of cooperation which enabled everybody to maintain a certain level of subsistence. While, undoubtedly, these practices have lessened considerably in Göreme over the past four decades, with many people either selling their gardens, vineyards, and orchards or no longer tending them so that they become unproductive, many people have held onto their inherited plots of land and, like Abbas, some have bought new gardens to add to their portfolio. Even whilst chasing wealth from tourism, many of Abbas's friends have tended to hold on to at least some of their gardens. Hence, simple binaries regarding positions, between peasant and entrepreneur for example, are not possible. Moreover, although some, for example Araghi (1995), suggest that the ongoing existence of 'the peasantry' is in question due to widespread processes of urbanisation and peasant economy dissolution under capitalism, others have argued for the importance of studying peasants-in-transition, or the new peasantry, and have thus pushed for a re-embracing of the term 'peasant' as an analytical concept.

In reference to contemporary Turkish society, Öztürk et al. (2018) argue that a new form of peasantry has emerged which 'takes on a somewhat amorphous quality, without clear divisions enabling us to say exactly who is and who is not a "peasant" and inviting the non-homogenous category of "semi-peasant"' (p. 253). Öztürk et al. (2018) suggest further that the new peasantry of Turkey is of particular interest in the way that it, through combining traditional practice with capitalistic entrepreneurship, both resists and exists because of neoliberal globalisation. This point resonates with Johnson's (2004) suggestion that contemporary peasant production is 'a particular form of production that is continuing because of, and in some cases despite, global capitalism' (2004, p. 63). Such paradoxical social forms inevitably occur in regard to other areas too. For instance, a resurgence of Islam in Turkey in recent decades has been argued to present a paradox in its coexistence with Turkey's simultaneous economic liberalisation processes, thus producing an emergence of economic Islam and 'Islamic capital' (Özbudun and Keyman, 2002). It is inevitable that any such entanglements manifest in different social and cultural contexts as peculiarly local, and perhaps hybrid, social dynamics. In Göreme, the particular way in which peasant continuities have created contradictions and fissures in what might otherwise be deemed globalised tourism business practices is interesting to consider.

My aim in this chapter, then, is to outline key *because of* and *despite* dynamics of tourism-related change, focusing in on aspects of peasant sociality and forms of production which have continued – and discontinued – as

Göreme's peasants-in-transition have become increasingly entangled with a globalised tourism economy. As I will go onto discuss, I see a core aspect of these peasants-in-transition dynamics of change as being the transition from beliefs about tourism being a *limited good* to it being seen as *unlimited good*. As a starting point in these considerations, I will next outline portraits of six people whose stories act as signposts for discussion as the chapter continues. These include revisiting some key characters who were depicted in portraits in *Living with Tourism (1)*, some of whom had initiated tourism business in Göreme. I will begin with Abbas who featured prominently in that first book (see esp. pages 17, 70–71, and 93 of Tucker, 2003).

Abbas

Abbas was born in 1955 in a cave-house in the upper parts of Göreme's Gaferli neighbourhood. His family lived in the cave-house until, when Abbas was still a young boy, they were re-housed into a small concrete '*afet*' (the government's disaster relief programme) house in the lower section of the village because of the danger of rock falling from behind their old house (the ridge now known as 'Sunset Viewpoint' sits above where the family's old house was located). For some decades after their relocation, the family continued to use their old house's cave-rooms for food processing and storage since the small, concrete *afet* house was not adequately set up for such practices. In the early 2000s, the old cave-house was sold and was later turned into a hotel.

The extent of Abbas's formal education was completing five years of primary school in the village; his family were too poor, he told me, for him to go onto intermediate school. I had got to know Abbas's elderly mother during my 1990s fieldwork (see portrait of "Anne" – meaning mother or mum in Turkish – in *Living with Tourism (1)*, p. 70), and she often told me stories of how hard life used to be for them and most other families in Göreme. She talked of bringing her five children up largely on her own because her husband, Abbas's father, was unable to make a living in the village and so lived away on the mountains tending sheep. She also told me that it was having to cook and sew by the light of an oil lamp within the dark cave of their old house that had made her go blind. While life for Abbas and his four siblings has not been altogether easy either, the opportunities – for the men at least – to make an income from tourism helped to improve their lot considerably when compared with their parents' generation.

Abbas prides himself on the fact that he opened the first tourist restaurant in Göreme. His brothers also were tourism pioneers; his older brother developed one of the first tourist horse ranches in Göreme, and his younger brother was the first Göremeli man to marry a 'tourist bride', from Scotland, together with whom he owned and ran one of the earlier – and most successful – pansiyons for backpacker tourists. As a result of his younger brother's marriage, Abbas has two nieces and a nephew who are bilingual, half-Göremeli and half-British.

As teenagers, Abbas and his friends had sold postcards to the few tourists who trickled through the village to visit the Göreme Open-Air Museum. After completing his military service in Cyprus in the mid-1970s, upon returning, he married, and, in 1977, he opened a small, simple restaurant called Maçan Restaurant (Maçan was the name of the village before it was renamed Göreme in the 1980s). For 18 years, Abbas successively opened and ran restaurants on the main street of Göreme (which changed name and location according to where he was able to lease a suitable building), until, in the mid-1990s, he decided to open a tour agency in partnership with his younger brother. The tour agency sold daily mini-bus tours around Cappadocia, as well as hiring out mopeds and bicycles. Along with working in partnership with his brother and wife's pansiyon to sell tours to the backpacker guests staying there, a grouping together with other tour agencies owned by Abbas's friends often occurred so that, when there were not enough customers to fill a mini-bus, they would share customers around. Abbas thus had legitimate reason to frequently drive around the village on his motorbike visiting his friends' agencies for a tea and business talk. He also, when necessary, helped his wife Senem tend their gardens and vineyards, some of which were inherited family plots while others were purchased using Abbas's tourism business earnings.

By the early 2000s, Abbas had paid national social security payments for long enough that he would receive a state pension. He sold the tour agency and used the sale money to extend and renovate his house, which was the *afet* house his family had been relocated to by the government in the 1960s. Moving into the new 'modern' house upstairs enabled Abbas to let the original ground floor house, and the rent received supplemented Abbas's pension in paying household bills and seeing their son through high school. After retiring from tourism business, Abbas also became more interested in his gardens. Experimenting with organics and different framing techniques for his grapevines, produce from the gardens became more plentiful so that, along with having very little need to buy food from the market, he and Senem were able to sell some fresh produce as well as preserved products they made from it. The high quality of Abbas and Senem's *pekmez* (grape molasses – see Chapter 7) became well-known, and each year, they would sell it to friends, neighbours, hotels, and restaurants.

Figure 2.1 Abbas and Senem harvesting produce in garden.

Although Abbas did not particularly want his son Emre to fall into tourism work, after leaving school, Emre got a job selling ice cream in the Göreme Tourism Development Cooperative shop at the entrance of the Göreme Open-Air Museum. Along with this job, throughout 2018–19, Emre helped his father in the evenings, filling a large mobile tank with water from the well Abbas had dug next to his house and towing it by tractor up the hill to hotels needing to buy extra water. The excessive use of water by the increasing number of hotels in Göreme often led to severe water shortages, especially during the summer months. Hotels in the upper parts of the old mahalles, where piped water was frequently cut off, started buying water from Abbas, whose house was situated in the lower part of Göreme and had access to underground water. Sometimes Emre's water deliveries were to the large new hotel of Abbas's good friend Osman, introduced next in this section.

Abbas often reflected during our conversations in recent years about how, while he was key in starting tourism in Göreme, it was others – some of them his good friends – who had become wealthy from tourism. Nonetheless, he also always maintained that he and his family were "comfortable" (*rahat*); able to keep up a modest lifestyle from his state pension, garden produce, and rent-money, he said that they had "enough" (*yeter*).

Osman

Being close in age and growing up in the same neighbourhood (*mahalle*), Osman and Abbas were friends from childhood. Like Abbas, Osman sold postcards to tourists as a teenager and did his military service in Cyprus.[1] When he returned from Cyprus, he got a job in a factory in a nearby town and later worked in the Göreme municipality office. In the mid-1980s, he decided to open a small tour agency in Göreme. He told me that business was good in the beginning because his tour agency was among the first in Göreme. However, when more agencies opened, his business went down, and so he decided to convert part of his house into a hotel in order to supplement the modest income he received from the agency. A little later, in the mid-1990s, he extended his business portfolio again by buying and renovating a cave-pansiyon. When I knew Osman during my 1990s fieldwork, the agency, hotel, and cave-pansiyon were the three main businesses he owned and ran, plus he had partnerships in other land and building developments in the centre of Göreme.

In the late 1990s, Osman had purchased an old cave-house, which had a large piece of land around it, situated at the very top of Aydınlı Mahallesi (*mahallesi/mahalle* is the Turkish word for neighbourhood; see Chapter 5 for a discussion of change in Aydınlı Mahallesi). Due to its position at the back part of the village, the land was very cheap at the time that he bought it. He did nothing with it for several years since he was busy looking after his three main businesses while also attending to his son's and two daughters' education. His oldest daughter studied English language and, married with children, went to live in Kayseri and worked as a high-school English teacher. Osman's son studied English language and culture at university in Cyprus, while the younger daughter studied psychology at Kayseri University. This younger daughter, as a teenager, went to live for a time in Canada where she was hosted and cared for by a Canadian friend of Osman's whom he had earlier met through his travel agency business. Although Osman's elderly mother objected strongly to her granddaughter's going away, Osman persisted since he felt sure that spending time overseas would broaden his daughter's horizons (see Chapter 6).

In the early 2010s, Osman began developing the large property at the top of Aydınlı Mahallesi property, and, in 2019, it opened as a high-end 'boutique' hotel. That summer, when he met me at the gate to give me a tour of his new hotel, Osman gestured into the property and said "Look, I am a rich man!"; he then gave a chuckle, to let me know that he would not make such an exclamation in all seriousness. During the tour of the hotel, he showed me several opulent cave-suites, with private terraces and ensuite bathrooms with marble fittings and jacuzzi baths. The hotel had extensive grounds with grass lawns and flowerbeds (which had frequent need for the services of Emre and Abbas's water-delivery tractor) and a large central breakfast restaurant. Osman and I went into the restaurant and sat chatting while drinking Turkish coffee, served to us by a young Afghani man whom Osman said had worked for him for several years. Most of his other employees were women from Nevşehir, who cleaned and made up the rooms and sometimes helped out in the breakfast room. In 2019, although some sections of the hotel were still to be completed, the hotel already had a steady clientele through *booking.com* and other online booking platforms. Osman said that he had put a lot of money into the hotel project and, at times, had needed financial help, for example, when – due to the 2016 attempted coup and associated political unrest in Turkey – tourism collapsed for a year and a half and he had virtually no income from his other tourism businesses. He told me that some of his friends – other hoteliers in the village – had helped him out. He then added that, with his new hotel having 24 rooms and suites selling for between 90 and 300 euros each, it would not take too long to pay off his debts. (However, the Covid-19 pandemic was later to significantly knock this plan).

Clearly proud of his new hotel, Osman said that, from being the first Göremeli man to open a tour agency business, this hotel was the "end result"; "I've always dreamt of owning such a place", he said. His son managed the hotel, while his wife ran the few rooms they still let in the original hotel above their house. Osman's wife also made jams and other traditionally made preserves for the breakfast buffets at the two hotels. Osman told me that, because most tourists these days book through online booking agencies and are able to write a review of their hotel after their stay, "we have to give good service, good breakfast and so on, because we have to maintain our impeccable rating". He went on to say that tourism is completely different now compared with how it used to be; "It used to be much closer between the tourists and us. Now they want a different kind of service from before". He said that in some ways, the online review system makes it more difficult to run a hotel, while on the other hand, the online booking platforms make things more equal among the hotels:

> Before, there were no tourist reviews, but instead we had to put up with the bloody *Lonely Planet*. I hated it. The writer just had her favourites and wrote them in the book and not anyone else. So the *Lonely Planet* made some Göreme people very rich and others had no chance at all. It was very unfair. At least everyone has an equal chance now.

As we stood talking on one of the terraces of the hotel, suddenly a young Asian woman came out of a room and climbed the steps going up to the balloon-viewing terrace. Osman said: "Look, pretty girls just pop up like that now; I was born too early! The young have it all here now – it is easy for them!" I asked if they realise they have it easy, and he answered that some do and some do not.

Hanife and Hayriye

Hanife and Hayriye, two sisters, are proud that they were the first Göremeli women to do business in the central part of Göreme. They opened their shop called 'Authentic Crafts' on the main street in the late 2000s, selling handcrafted scarves, necklaces, and other beaded and crocheted jewellery. Their business endeavours had begun a few years prior to opening the shop when, in 2003, they had set up a simple stall by their family's garden in a valley close to Göreme known as 'Love Valley' where tourists would often go for a walk. Prior to that, one of the sisters had worked as a cleaner in hotels. They said that, before, women could only do that kind of work.

Figure 2.2 Hayriye at valley stall.

When they first opened the stall in the valley, they would sneak out of the house telling their father they were going to work at the garden. "Because we were hidden in the valley, nobody knew what we were doing", they said; "Our father would never have given us permission to go there and do a stall". Gradually, however, their father, and later their husbands, accepted their business activities because they were bringing in money; "I haven't needed to ask for money for 7 years!", Hayriye told me in 2010.

When they first opened the shop on the main street, the sisters felt extremely shy, they told me, and hid behind the shop counter whenever a local man passed by. However, due to their bravery in opening the shop, other women felt they were able to come to the çarşa (central streets); "It doesn't cause gossip anymore", they said, "now there are lots of women in the centre – it has changed very quickly". I had that particular conversation with the two sisters in 2010 and, as they suggested, in the space of just a few years, there had been a notable increase in Göreme women's presence in the central streets (see Chapter 6 for further discussion of these changes).

Even after opening the shop, for many years, the sisters continued to run the stall in the valley, with usually Hayriye going to the valley while Hanife looked after the shop. They said it was worth keeping the stall because, as well as backpackers going hiking in the valleys, tour groups were often taken there to photograph the fairy chimneys. Hayriye did well selling to these groups, she told me, because as well as selling necklaces and scarves, she also sold freshly squeezed orange and pomegranate juice. They told me in 2010: "The stall sells more than the shop, so that's why we keep the stall in the valley. Also, now our father sometimes comes and visits the garden – and of course the stall!" In 2016, they closed the stall, however, because following the attempted coup in Turkey, tourism dropped off and there were no tourists in the valley. The sisters decided, at that point, to keep only the shop going.

Over the years of developing their business, as well as selling directly to tourists, they regularly secured commissions for the products from individuals or companies overseas (including in Spain, Greece, the Netherlands, Japan, and Kuwait). Consequently, they set up women's cooperatives in other Cappadocian villages in order to produce the handiwork items needed to supply such commissions, as well as to supply stock for the stall and shop. In 2010, they had 40 women from other villages working for them, and in 2019, this had increased to 60 women. Hanife and Hayriye told me that they were pleased to be able to help out those poorer women by paying them well for their work and by giving them work they *could do* – even those in the more conservative villages – because they were able to do the work from their own homes.

Overall, not only are Hanife and Hayriye good examples of success in business in Göreme, but they were at the forefront of significant change in Göreme whereby women were able to enter the tourism business space. They told me that their success was largely due to their doing business *together* – the *two* of them – because they "made each other brave". They added that, because of them, other women became brave too.

Mustafa

Mustafa[2] is approximately 15 years younger than Abbas and Osman, and when I knew him during my 1990s fieldwork, he did not yet have his own tourism business. Having left school at age 11, he had started working in tourism by cooking at a campsite near the Göreme Open-Air Museum where backpackers travelling on 'Top Deck' company buses stayed – hence, giving Mustafa his nickname "Top Deck". He did his military service near the Turkish-Iraq border during the Gulf War, and after he returned, he worked in various friends' pansiyons, cooking and 'hosting', among other general pansiyon jobs. He worked as ground-crew for Göreme's first hot-air ballooning company for a few years during the late 1990s, and after that, throughout the 2000s, he was the chef in a relative's café-restaurant on the main street.

In the 1990s, Mustafa married a 'tourist' – from South Africa – he had met in the pansiyon he was working in and they went on to have two daughters together. They lived with Mustafa's elderly mother in the old cave-house which had been left to Mustafa by his father. In this family's case, when their father died, Mustafa inherited the house while the family gardens were passed on to Mustafa's older brother. One of those gardens, having an impressive setting on the edge of the village and a Byzantine cave-church in a neighbouring fairy chimney, was turned by Mustafa's brother and sister-in-law into a small tourist venture serving tea and small snacks to tourists who came to see the church. Reportedly, one day, during the 2000s, Helen Clark, who at the time was the Prime Minister of New Zealand, came to visit the cave-church and had a glass of tea. Mustafa's sister-in-law was unaware that she was serving tea to a Prime Minister and gave her the same warm welcoming hug she gave to all her female guests.

Mustafa's own tourism business venture began in the late 2000s when, having inherited the old family house, he renovated the large cave situated beneath the house and converted it into a restaurant. As the cave had originally been the stable where the family's donkeys and horses were kept, the renovation involved substantial cleaning, but the end result – decorated with Turkish rugs, cushions, and lamps –

was an atmospheric cave-restaurant in which Mustafa cooked and served a simple menu of "local dishes". Reviews of the restaurant on the *Tripadvisor* platform were consistently favourable, many saying that Top Deck's is the best restaurant in Cappadocia, and in 2015, it was ranked "No. 1" on *Tripadvisor*. While pleased with this ranking, Mustafa complained that it almost made him "too busy"; he said that the restaurant was so popular that they were fully booked every night and he even had bookings from people six months in advance of their trip to Göreme, made to ensure they got a table. He also said that "because I'm No. 1, people expect No. 1 standard, so there is constant pressure to keep it up"; "I wish they would write bad reviews about me for a change", he joked, although there was likely more than a grain of seriousness in his words. Despite the restaurant's popularity, Mustafa told me he does not raise its prices, saying that he always wants to be reasonably priced and not over-charge: "If someone comes all the way from New Zealand, for example, I shouldn't over-charge them, because they chose to come here and so I should give them my hospitality", he told me.

The volatility of tourism meant that times would come when Mustafa's, as with other businesses, would not be so busy. Following the attempted coup in July 2016, for example, almost no international tourists came to Turkey right through to the start of 2018, and businesses in Göreme had to rely on the few Turkish tourists travelling around. Things picked up again in the summer of 2018, however, and Mustafa's restaurant once again became fully booked most evenings. During the summer months, his daughters helped out in the restaurant doing table service. The rest of the year, they were away at university. Mustafa told me that it was important to him that his daughters had a good education: "Then they have an open future", he said; "They can take over my restaurant, or not, they choose, I want them to have an open future". Besides his daughters helping in the restaurant during the summer, like Osman, Mustafa employed young Afghani men, and he too told me that they were good workers. He had trained two of his Afghani employees how to cook according to his recipes and methods so that, when he wanted to "escape" from his busy business by going out of the village to spend time at his recently built 'garden-house', they could take over in the kitchen and prepare the food for the evening's service. He told me that he was slightly worried that he had trained them so well that they might soon go off and open their own restaurant. In some ways, however, he did not worry about this, since he was "tired of tourism business", he told me, and having built the small house at his garden, he enjoyed more and more spending time there to escape the busy-ness of Göreme.

Mehmet

Mehmet is slightly younger than Mustafa, and when I met him during my 1990s fieldwork, he worked as a salesman in the carpet shop owned by his father. He developed good relationships with an international array of the shop's tourist customers and, in the winter, when there were no tourists in Göreme, he often travelled to visit international friends, particularly in the United States, and would sell carpets through their networks. During the early 2000s, he bought a few old, crumbled cave-houses near the top of Aydınlı Mahallesi (not too far from Osman's later hotel development) and, with the financial help and know-how of a successful hotelier friend, he conducted extensive re-builds and renovations to develop a 'boutique' cave-hotel and restaurant. He continued to grow the size of his hotel as neighbouring cave-houses came up for sale and, together with his business partner, also opened a travel agency and a hot-air balloon company.

One of Mehmet's business strengths was his recognition that, if he travelled in the world, he would increase his own understanding of where his customers came from. He frequently travelled around the world; as well as to the United States, he went to Australia, New Zealand, China, and several countries in South America. This allowed him to better understand current and potential tourist markets as well as, he believed, to more effectively 'host' his hotel's 'guests'. While many Göremeli people – men in particular – developed a certain cosmopolitanism from their tourism involvements (see Chapter 7), Mehmet always struck me as developing an extensive understanding of the global world of travel. For example, he recognised earlier than others the potential of the Chinese young-independent tourist market and, in 2018–19, he worked with social media to such an extent that he developed a role for himself as a key 'influencer' for Cappadocia, and Göreme in particular.

Following the dramatic drop in tourist numbers during 2016–17 following the mid-2016 attempted coup, Mehmet energetically resumed his promoting of Göreme on social media by inviting international influencer-bloggers to stay in his hotel, go hot-air ballooning and to put the message out that "it is safe here" again. Aware that American, European, and Chinese tourists were still rather nervous about travelling to Turkey, he actively connected with the Russian and Ukrainian markets, inviting models, bloggers, and social media travel influencers from those countries to stay in his hotel. It was at Mehmet's hotel that the idea was created to lay colourful Turkish carpets and a photogenic "breakfast table" out on the hotel's roof terrace for the models and influencers to do social media photo-shoots while having breakfast with the hot-air balloons flying in the background. This practice was soon copied by other hotels and, quickly gaining momentum, became a highly influential craze (see Chapters 3 and 4 for discussion on *Instagram balloons* and hotel *Instagram* terraces) which was central in Göreme's becoming a global "bucket list" item as well as being named on *Tripadvisor*'s top ten list of places to experience in 2019.

By 2019, Mehmet had over 200,000 followers on *Instagram*, plus he tagged pictures taken on his hotel terraces with balloons in the background on sites such as 'Beautiful Hotels' and 'Beautiful Places' which, he told me, had 30 million followers. His proactive social media work led to him being invited to Istanbul to receive the 2019 'Boost Your Business' Award for best *Facebook* and *Instagram* usage for business in Turkey. Not only had he learned how to use social media to his own business advantage, but his active posting and tagging, as well as inviting and hosting hundreds of blogger-influencers, promoted Göreme and Cappadocia in general alongside his own hotels and hot-air ballooning company. Importantly, also, through this process, Mehmet learned that social media enabled a business, and a place, to seek and create new tourist markets; in other words, he came to understand tourist markets as *growable*, or extendable, something which had not been the view of Göreme's earlier tourism entrepreneurs.

Since I had first known him 20 years earlier, by 2019, Mehmet had become a wealthy man. He had married a woman who had grown up in Ankara but whose family had originated from Göreme, and they lived with their two young children in a large but not overly ostentatious house they had designed and built in the *Yeni (New) Mahallesi*, which used to be known as the *afet* (disaster)-house neighbourhood, where Abbas lives. He continued to buy cave-houses in the old neighbourhoods to convert into hotels or to add to his already-existing hotels, and hence would gradually own whole sections of a street, just like owning a colour group on a *Monopoly* board. He also bought and renovated the 'Konak House', an Ottoman mansion where the local 'paşa' (noble) once resided, and having had the decorative frescos in the mainrooms painstakingly restored, he used the impressive building to host business associates as well as allowing bloggers to hold photo-shoots in the frescoed rooms. Between his multiple boutique hotels, the Konak House, the travel agency, and hot-air ballooning company, Mehmet had become the '*patron*' (boss) of a great many employees by 2019. The majority of these were women and men from nearby towns who sought employment, such as cleaning and kitchen work, in Göreme's tourism businesses. Others, usually those in the more managerial positions, were relatives or friends of Mehmet's. He had also hired various foreign nationals in his businesses for certain strategic reasons: a Chinese student had worked in the office and reception of his hotel so that she could act as cultural mediator and promote the hotel on Chinese social media; a Russian social media 'influencer' was employed as his '*Instagram* Manager'; and an Australian woman and another person from Belgium were employed in the offices of his travel agency and hotel. Since his humble beginnings in business 20 years earlier, Mehmet not only became highly successful in business himself, but, it can be said, he was key in influencing Göreme's world-wide social media fame.

Göreme, tourism, and 'peasant' ('*köylü*') (re)production

That Abbas and his friends contended that Abbas could not be *köylü* (peasant/villager) because Göreme was no longer a *köy* (village) is suggestive of a fundamental shift in the nature of the place to which they belong. In *Living with Tourism (1)*, I had pointed out that while Göreme is technically a *kasaba*[3] which is the Turkish term for a place between the size of a village (*köy)* and a town (şehir), 'the people of Göreme themselves usually refer to the place as their village and to themselves as villagers (*köylüler*, which also translates as peasants)' (Tucker, 2003, p. 22). In rural Turkish society, the village (*köy*) was understood to be the key unit of sociality. As Paul Stirling (1965, p. 29) put it in his Turkish village ethnography: 'People belong to their village in a way that they belong to no other social group. On any definition of a community, the village is a community'. Likewise, Carol Delaney (1991, p. 202) said in relation to Turkish village society that, 'more than merely a place one lives, a village is an indelible part of one's being and identity'. Hence, that Göreme is "no longer a köy" would appear to strike at the heart of Abbas's and his friends' being and identity.[4]

It is possible, also, that denial of Abbas's return to being *köylü* was related to negative associations of the word – similarly, perhaps, to the everyday derisory connotations of the terms 'peasant' and 'peasantry' in English.[5] In other words, Abbas and his friends may be keen to deflect a longstanding 'particular view of peasants, namely that they are conservative, uninterested in change, and unintelligent' which Delaney (1991, p. 220) observed to be the prevalent view of Turkish villagers among urbanites. Yet, Öztürk et al. (2018) conjure an altogether more positive view, 'emphasising the adaptive and creative agency of the peasantry' (p. 244), and suggesting that this agency manifests as 'a kind of resistance, which involves maintaining the peasant tradition of self-sufficiency and abstaining from commoditisation of the farm and social relations' (p. 245). Indeed, Abbas would appear very much to display this kind of adaptive resistance and so, although he preferred to deny his status as *köylü*, I would contend that he – along with a great many other Göremeli people – continue to share aspects of sociality and being which could be described as peasant characteristics. Moreover, as Öztürk et al. (2018) suggest, it would appear that such peasant continuities, while increasingly loaded with ambivalence, act in many ways as forms of resistance to 'the neoliberal squeeze of capital' (ibid., p. 245).

It is interesting to consider, therefore, how various strands of Abbas's and others' peasant sociality have continued and become entangled within tourism development efforts during the past four decades. Of particular relevance here are, firstly, what June Nash (2007) refers to as 'peasant notions of scarcity' (p. 36), and secondly, but relatedly, what Van der Ploeg (2008) describes as the peasantry's 'strong preference for regulating relations within the community through social mechanisms, norms and values – that is, through non-commodity relationships' (p. 140). In regard to notions of scarcity, peasant sociality is broadly considered to engender a form of moral economy borne principally out of what George Foster (1965) originally identified as the peasant 'image of limited good' belief. This is the belief, or worldview, that:

> all of the desired things in life, such as land, wealth, health, friendship and love, manliness and honour, respect and status, power and influence, security and safety, *exist in finite quantity* and *are always in short supply.*
> (Foster, 1965, p. 295)

Foster (1965) saw this belief – that 'there is no way directly within peasant power to increase the available quantities' (p. 295) – as a cognitive orientation which helped explain many peasant community behaviours. As I will go onto explain, such peasant community behaviours were clearly apparent in the way

Figure 2.3 Göreme village and surrounding gardens, 1912. Photograph courtesy of Göreme Tourism Development Cooperative.

in which tourism business developed in Göreme, especially during its initial two decades.

Premised on the idea that 'a peasant sees his existence as determined and limited by the natural and social resources of his village and his immediate area' (Foster, 1965, p. 296), Foster himself acknowledged that the notion of limited good infers a more or less closed system, even though communities are seldom such. Certainly, the 'new peasantry', or 'peasants-in-transition', operate in anything but closed systems, fully integrated as they are into national and global capitalistic relations. However, in central Turkey 'the village was often the only life-world' (Keyder, 1993, p. 171) until the 1950s and 1960s saw an opening up due to 'market adaptation, new inputs and changing technology ... and, of course, the beginnings of urbanisation' (ibid.). In the 1960s, also, extensive migration programmes enabled substantial numbers of central Anatolian villagers to migrate to northern Europe. In line with these changes, during the 1960s and 1970s many Göreme men and families migrated either to urban areas of Turkey or to northern Europe, in particular, Germany, Belgium, and Holland. A significant number, also, having bought trucks using money earnt in or sent back from Europe, worked in haulage transporting goods between Europe and the Middle East. Those who stayed closer to home sought cash-earning opportunities in nearby towns. Hence, as with other villages in central Turkey, Göreme was far from being a closed system as it went through the second half of the twentieth century, and so the villagers could already be considered 'peasants-in-transition' even prior to the start of tourism in the area.

Nonetheless, many aspects of peasant sociality premised on image of limited good beliefs were observable in Göreme's initial tourism developments, such that tourism there could be said to have developed in ways which were aligned with 'peasant' forms of (re)production. Paradoxically, these developments occurred within a wider context of the Turkish economy's embracing of neoliberal capitalism in the 1980s. A low level of international tourists were already visiting Turkey prior to 1980, and some of those trickled through Cappadocia visiting the rock churches particularly around the area later enclosed as the Göreme Open-Air Museum. These tourists were whom Abbas and Osman sold postcards to and who may have visited Göreme village's first restaurant, opened by Abbas in 1977. It was in the 1980s, during a period of relative stability following a military coup at the start of that decade, that tourism development began apace. Spurred on by economic liberalisation policies and legislation, Cappadocia was earmarked as a national centre of 'cultural tourism development'. As described in Chapter 1, however, Göreme's position in the centre of the Göreme National Park meant that the mass tourism developments occurring elsewhere, and sometimes very close by, did not occur in the village itself. Indeed, throughout the 1980s, despite the name of the village being changed to Göreme (from its previous name of Avcılar and prior to that Maçan) so that it would become more directly associated with the nearby Göreme Open-Air Museum site, Göreme struggled to become known as a place with any tourism facilities. While the town of Ürgüp, ten kilometres from Göreme, developed a reputation as the main tourism centre of Cappadocia and experienced an influx of foreign and national hotel chains constructing three-, four-, and five-star hotels, the village of Göreme became known as a place only for backpacker tourists, and "cheap" ones at that.

However, it was this reputation as a backpacker place which enabled locally owned small and micro-businesses to be developed there. Moreover, Göreme's protected position within the national park, preventing the large hotel companies from moving in, was arguably what enabled Göreme's backpacker tourism industry to thrive.[6] A defining feature of backpacker tourists was understood to be their desire to differentiate, and separate, themselves from whomever they considered to be 'mass tourists' (Tucker, 2003). Being based in Göreme allowed backpackers to separate themselves in this way since it had no large hotels for such 'mass tourists'. Moreover, Göreme's featuring in the main backpacker guidebooks as having plenty of lower-budget accommodation and services depicted the village as an ideal location for backpackers to base their stay in Cappadocia. By the end of the 1980s, there were 50 or so 'pansiyons' – small accommodation businesses for backpackers – in Göreme, and by the end of the 1990s, this figure had gone up to around 60.

Importantly, the fact that the backpackers sought lower-budget accommodation and services enabled Göremeli villagers to develop businesses with little capital input. Often, for example, villagers started out by converting one or two rooms in their cave-home for tourist accommodation, adding to their business incrementally only as money for decoration and furnishings allowed. As one entrepreneur told me during the late 1990s: "The people are just doing their job and their own businesses within their own possibilities, rather than getting credit from the bank or things like that – some do but very few" (quoted in Tucker, 2003, p. 97). This idea of the villagers creating "their own businesses within their own possibilities" very much aligns with what are generally regarded to be key characteristics of peasant (re)production practices. For example, Van der Ploeg (2008) noted a logic of survival, as well as cooperation for survival, as key features of peasant production based on notions of scarcity – or limited good – engendering a symbiotic unity between production and consumption. Similarly, according to Johnson (2004), peasant production is characterised both by a logic of subsistence and the maintenance of significant control over the means of production. In Göreme's tourism context, the developing backpacker tourism suited adaptation of peasant forms of production so that tourism entrepreneurship developed *within* peasant sociality, as a hybrid incorporation of peasant practice and based on a logic of survival. Moreover, with high levels of owner-operator businesses and limited use of credit, full ownership and control over the means of production were able to be maintained.

Van der Ploeg (2008) also pointed out that the logic of survival may be constantly adapted to varying circumstances and types of production – such as, in this case, tourism business, and therefore does not necessarily relate to subsistence in the sense of self-provisioning of food. Nonetheless, most Göreme households held onto their gardens – even if household members were involved in successfully developing tourism businesses. In one sense, this might be explained in what Öztürk et al. (2018) suggest as Turkey's new peasantry's continuing 'to value their land and farms and communities for their own sake, as ends in themselves, maintaining what we may call an *internal communing* ethos' (p. 251); hence the notion of (*re*)production. Besides this impetus of an 'internal communing ethos', Bryceson (2000) has pointed out that many 'peasants-in-transition' retain their land in order to maintain a form of 'subsistence fallback', hence 'safeguarding peasant survival in the face of adversity' (p. 312). In Göreme, following the initial heyday of backpacker tourism growth during the 1980s, the villagers were indeed faced with adversity as the next decade and a half saw a series of significant regional and global 'shocks'. The Gulf War in 1990–91 came as the initial shock and, causing a dramatic downturn in tourist numbers, it taught entrepreneurs important lessons both about the precarious nature of tourism business and that tourism was heavily dependent on external factors beyond their control. This understanding of the precarious nature of tourism business was fed further by ongoing regional and global disruptions to tourism, including terrorism in the southeast of Turkey, the war in former Yugoslavia, the global economic crisis of 1994 and the earthquakes in the northwest of Turkey in 1999. Tourism's global fragility was highlighted once again with the 9/11 attacks of 2001, followed by the allied invasion of Iraq. The sense that international tourists might never feel safe again in coming to Turkey served to remind Göremeli people not only of the limits of tourism 'good', but also of the need to safeguard survival through maintaining a form of subsistence fallback.

Further examples of adaptation of peasant forms of production into Göreme's early tourism developments included cooperative behaviour and what Van der Ploeg (2008) termed 'pluriactivity'. Cooperative behaviour occurred through pairs or groups of individuals creating tourism business partnerships (*ortaklık*) so as to share the capital costs and thereby the risks associated with opening a new business. Networks of both shared costs and reciprocity of labour are argued to be integral to peasant communities in that they constitute 'the moral economy of the peasant' (Scott, 1976). Such partnership systems were already a part of social life in Göreme prior to tourism development, just as they were known to be prevalent in relation to the cooperative aspects of Turkish village life in general (Delaney, 1993; Stirling, 1965). As tourism businesses were developed, many entrepreneurs became involved in multiple *ortaklık*, so that any individual could be involved in a variety of tourism services at any given time (see Mottiar and Tucker, 2007). This involvement in multiple businesses, or 'pluriactivity', could be seen as a further example of peasant strategy of survival in that it allowed some businesses to supplement others when necessary: 'Through engaging in pluriactivity, dependence on banking circuits and moneylenders can be avoided' (Van der Ploeg, 2008, p. 33).

The practice of having multiple *ortaklık* (partnerships) and hence involvement in several complementary businesses also linked to 'limited good' in that it was a way for Göremeli entrepreneurs to spread out their metaphorical 'net' in order to catch as many of the available tourists as possible. For instance, *pansiyon* owners often started up or took part ownership of a tour agency –

just as Abbas's brother did – so that tourists staying in the *pansiyon* would go on day tours with the linking tour agency. Similarly, some carpet shop owners went into partnership in a *pansiyon* business and a tour agency so that tourists patronising those businesses would automatically be sent – or taken – to that particular shop if they were interested in buying a carpet. Hence, Göreme entrepreneurs' engagement in multiple businesses through *ortaklık* networks aligned with cooperative peasant production practices linked both to an image of limited good and maintenance of autonomy and control in production.

A further example of peasant community cooperation was the forming of a tourism-related cooperative organisation – named the 'Göreme Tourism Development Cooperative' – in the late 1980s. The organisation engaged in a variety of tourism promotion activities, including the production of brochures and posters and distribution of these at nation-wide tourism fairs. The Cooperative also developed businesses, including a souvenir shop at the entrance of the Göreme Open-Air Museum and a general store in Göreme's centre, and facilities for villagers such as a teahouse and snooker hall for village men to gather in the winter months. With most Göreme households being members of the Cooperative, they in effect had joint ownership of these businesses and facilities. Once a year, each household received a dividend, usually in the form of a kitchen or other household appliance, and otherwise the profits from the businesses were used to develop further businesses (in the late 2000s, the Cooperative developed a Turkish bath, or *hamam*, and a shop complex on Göreme's main street). Importantly, the shop at the museum entrance – where Abbas's son got a job after leaving school – was pretty much the only way that the villagers could earn from the great many "group tour" tourists they saw daily going straight through the village in their buses heading for the Göreme Open-Air Museum approximately two kilometres away. Those were the tourists who stayed in the large hotels outside of the national park perimeter and, being bussed around between the main attractions, were seldom able to roam in Göreme village let alone patronise the businesses there.[7]

In contrast to the apparently plentiful (the museum car park always seemed to be full of tour buses) and yet inaccessible tourists on group tour, the backpacker tourists who did stay in the village always seemed to be too few in number, as well as being on a limited budget. Tourists and tourism thus became incorporated into the peasant *limited good* orientation; as Van der Ploeg (2008) remarked, 'the notion of plenitude is definitely at odds with the life worlds of peasants' (p. 43).

Tourists/tourism as limited good: competing to remain equal

> ...an individual or a family can improve a position only at the expense of others. Hence, an apparent relative improvement in someone's position with respect to any "Good" is viewed as a threat to the entire community.
>
> (Foster, 1965, p. 297)

That tourists seemed to *exist in finite quantity and always in short supply* had significant influence in the ways in which tourism business would play out as it continued to develop beyond the 1980s into the 1990s. Firstly, limited good beliefs engendered in Göreme's tourism business what Bailey (1971) identified in relation to peasant communities as a constant competing to remain equal: 'Skills and energies go into keeping people in the place that they have always been: they run hard in order to stand still. It is the kind of world that stamps heavily upon change and innovation' (Bailey, 1971, p. 23). This stamping on innovation for fear that the innovator might 'get ahead' of others concurs, also, with Stirling's (1965) noting of the 'proverbially conservative' character of Turkish villagers. Indeed, the early tourism entrepreneurs, such as Abbas who opened the first restaurant and Osman who was among the first to start a tour agency business in Göreme, were criticised and ridiculed by their peers. Abbas told me that other villagers would laugh at him and say 'What on earth are you doing that for? Who will come and eat there?' Abbas and Osman persevered, however: "I knew that tourists needed to eat, so I continued", Abbas went on to tell me. Seeing that they had some success, others soon imitated their innovations (see *Living with Tourism (1)*, Chapter 5, for more detail on these early business developments).

As tourism business developed and proliferated in Göreme, in line with the 'competing to remain equal' principle, the high levels of imitation meant that the businesses within each sector tended to be very similar to each other, with no single business able to maintain any significant point of difference. They copied each other's décor and menus, for example, and if one restaurant built a wooden terrace outside for customers' seating, the others followed so that the restaurants would all end up looking the same. Some earlier innovations were introduced by foreign incomers, such as the Scottish wife of Abbas's brother, who, having been a backpacker herself, had a good idea of what backpackers' desires and tastes were in relation to décor and accommodation services. As the businesses in which these 'tourist' incomers worked became kitted out in the desired rustic style with communal areas with floor cushions around low tables for backpackers to hang out and meet each other, soon the majority of pansiyons would be adapted with a similar look.

While such levelling mechanisms generally meant that no business would stay ahead of the game for long, some businesses had advantages that were more difficult to emulate or address. For example, Abbas's brother's pansiyon was considered to have an advantage in its Scottish 'host', who not only understood and could communicate well with her guests, but also could communicate well with, and befriend, the writer of the *Lonely Planet* guidebook. That pansiyon tended to feature at the top of the *Lonely Planet's* recommended 'budget accommodation' list over many editions. Such inequities, referred to by Osman in the portraits above, led to high levels of business 'jealousy' and could incite fierce criticism, gossip, and even fighting. Aligning with the *limited good* social principle that one person's gain with respect to any "good" must by necessity be another's loss, jealousy (*kıskançlık*) was often the subject of entrepreneurs' complaints about tourism's detrimental effects on villager relations and sociality.

With tourism good seen as limited and finite, and tourists always seemingly in short supply, business owners resorted to various measures aimed to bring down any business which appeared to have too much success, for example, when one or another restaurant appeared to be busier with more customers. As Bailey (1971, pp. 19–20) explained:

> People remain equal because each one believes that every other one is trying to better him, and in his efforts to protect himself, he makes sure that no one ever gets beyond the level of approved mediocrity. Equality, in communities like these, is in fact the product of everyone's belief that everyone else is striving to be more than equal.
>
> (1971, pp. 19–20)

An example of such dynamics amongst Göreme's tourism businesses occurred one summer when one of the restaurants on the main street hired a young Canadian tourist to help out. This incited fierce jealousy among the neighbouring restauranteurs as they assumed the Canadian was touting and attracting 'too many' customers to that particular restaurant. Whilst one strategy may have been for the neighbours to follow suit and employ a tourist tout in their own restaurants, instead they called the *jandarma* (from French word *gendarmerie*, meaning police) and reported that a foreigner was being illegally employed nearby. From then on, since the *jandarma* were now alerted, none of the restaurants could obtain the casual help of a foreigner.

Hence, at times, competition tactics used by one business owner to harm another's business may also cause harm to his own business in some way, or damage his own profits. A further example is in an anecdote I previously told in *Living with Tourism (1)* involving a *Göremeli* businessman who owned a hotel next door to the original hot-air balloon company office. The owners of the balloon company were wondering why they never received customers recommended by the hotel situated very close by. Indeed, they learned that the hotel was not recommending ballooning to its guests at all, but rather was putting them off from ballooning by telling them it was dangerous. The balloon company owners could not fathom the situation because, in their view, the hotel owner only had things to gain by sending his guests ballooning, not least the hefty commission he would receive. From a discussion of the situation with their Göremeli employees, the balloon company owners deduced that the owner of the hotel was discouraging his guests from ballooning for a mix of two reasons: firstly, he assumed his guests' *limited* budget would only stretch to one big-ticket item and he would rather recommend them to buy a Turkish carpet, and secondly, he was jealous of the visible success of the balloon company. Even though there was no direct business competition between them, his jealousy meant that he wanted to bring their success down, even if it meant him receiving no commission.

The commission practice referred to in the story above was a further iteration of the 'limited good' business practices creating a web of business relations, although a less formal network than the *ortaklık* business relations discussed above. While commission payments were in themselves referred to as *"haram para"* (dirty money) by villagers, it became accepted during the 1990s, as a normal part of tourism business, that commission could be earned by the owners and workers of guest houses, restaurants, and tour agencies when making recommendations to their own customers to do a particular tour or activity. Initially, 'Turkish carpet'-shopping was the main business sector engaging a commission system, mainly involving pansiyon owners taking guests to the carpet shop of their "friend" or "relative". The practice gradually became more widespread so that anyone – such as a waiter in a café or indeed anyone tourists just happened to meet – could try to take tourists to a specific shop or agency with the aim of receiving commission on what they purchased. Later in the 1990s, the practice became embroiled in the hot-air ballooning business too (see the *"Money-maker Balloons"* sections in Chapter 3).[8] Again, tensions in commission practice were compounded by the 'limited good' notion that the number of tourists coming to the village was fixed, so that one entrepreneur's gain would be viewed as another's loss. The earning possibilities from commission thus led to an increased fervour in business competition and in the sense of villager entrepreneurs, especially accommodation business owners, 'owning' their tourist guests. There were often ferocious arguments where owners of pansiyons accused tour agency workers or waiters from restaurants of stealing *"their"* guests by getting commission for a recommendation they made.

Due to the strong desire in everyone 'that no one ever gets beyond the level of approved mediocrity' (Bailey, 1971, p. 19) – that is, that no one should become more successful than everybody else – frequent price wars occurred whereby the businesses would constantly undercut each other's prices so that, in the end, no one was making money. To help alleviate this, in the mid-1990s, business owners formed two associations, one amongst the accommodation owners and the other for tour agency owners. These associations encouraged cooperation amongst businesses in various ways, but their most significant role was the setting of minimum prices. Each year at the start of the tourism season, the Accommodation Association members would come together and agree on a minimum price for *pansiyon* beds that year. The Tour Agency Association did the same for their day-tour products, with each member signing an accord agreeing that any member who breaks the minimum price code was punishable by a large fine. Based on a self-recognised readiness to give in to tourist bargaining, as well as the tendency towards 'competing to remain equal', this was a mechanism organised cooperatively by entrepreneurs to prevent *themselves* from under-charging. It thereby also challenged tourists' position of power (to bargain down prices).

The Accommodation Association enacted a further levelling mechanism through its setting up of an 'Accommodation Office' in the bus station. In this small building, each *pansiyon* could place an advertisement for newly arriving backpackers to take a look at all of Göreme's accommodation offerings and choose where they wished to stay. During the 1980s, there had been

Figure 2.4 Advertisements in the Accommodation Office.

frequent problems with *pansiyon* touts aggressively vying for attention as tourists stepped off their bus upon arrival into Göreme. After the Accommodation Office was set up, such touting was prohibited by the Accommodation Association so that arriving tourists would be politely invited to go into the Accommodation Office and choose a *pansiyon*. As well as alleviating the touting problems, it was hoped that this cooperative marketing technique would remove some of the inequality created by the selective advertising in the backpacker guidebooks, such as the *Lonely Planet* as Osman mentioned, by giving equal advertising space to each establishment. However, while it did serve as a levelling mechanism to some degree, most backpackers had made their choice of accommodation from looking at their guidebook before they arrived. The Tour Agency Association similarly created cooperation among the different agencies whereby they 'shared' customers in order to ensure a full load in each agency's mini-bus. Since all of the day-tour products were identical – due to copying and price-levelling mechanisms – the sharing of customers was easy to do.

So, whilst the peasant tendency towards 'competing to remain equal' gave rise to fierce competition tactics among businesses, there was simultaneously significant cooperation between business owners; indeed, the dynamics among all of Göreme's tourism businesses during those first two decades was a curious mix of both cooperation and competition. Van der Ploeg (2008) describes the peasant cooperation orientation as 'almost always a well-cared for balance between the individual and the collective' (p. 34), whereby cooperation does not negate individual interests since mutual arrangements of cooperation and support function as a 'safety belt' in times of difficulty. Moreover, cooperatives strengthen the functioning and autonomy of the wider group, as with the case of minimum price-setting serving to resist tourists' bargaining attempts, and so it may be precisely 'through cooperation that individual interests are defended' (ibid.).

Personal relations and histories also need to be factored into any tension that may exist between the two opposing tendencies to compete and to cooperate. In Göreme, certainly, tourism business relations have always been such that those who were competing against each other in business were likely at the same time to be friends, neighbours, or even relatives. As mentioned in the portraits above, Abbas and Osman had been friends since childhood, and so while they were in competition with each other when they both owned tour agencies, they also would pop around to visit each other to drink tea, chat about goings-on, and perhaps play a game of backgammon. In *Living with Tourism (1)*, I described relations between village men competing in tourism business as being akin to elastic; 'at times they stretch apart from each other, and then they bounce back, but they are constantly connected' (Tucker, 2003, p. 113). As one Göremeli business owner told me:

> I am still very good friends with my business rivals. We always visit each other and have tea and talk. But, on the other hand, tourism came and so you have to take a part of it, like a cake…. It is really becoming very hard, being friends for many years, but … you must live somehow, and some days you may disturb him and other days he may disturb you.

While the continuation of peasant *limited good* beliefs could be seen as the cause of the many and frequent outbursts of jealousy and ferocious bickering among business owners, the peasant mode of production could also be argued to follow along the lines of 'sustainable' tourism at the local, or community, level. Based on an image of limited good and 'competing to remain equal' principles, the peasant mode of (tourism) production in Göreme meant that sharing and distribution of benefits occurred and the community remained somewhat, even if not entirely, equal. Moreover, levels of local ownership, local decision-making, and local community economic benefits were extremely high as tourism developed in Göreme, especially during its first two decades.

A comprehensive survey of tourism business ownership in Göreme conducted in the mid-1990s by Turkish doctoral researcher Cemil Bezmen (1996) showed that out of a total of 188 tourism-related businesses in Göreme at that time, 161 were locally owned (by Göreme men). This was in contrast to the nearby town of Ürgüp, where foreign and non-local investments leading to loss of ownership and autonomy at the local level were argued by Tosun (2001, p. 295) to 'contradict principles of sustainable tourism development' (Tosun, 2001, p. 295). A key difference in what we might call the 'local sustainability' trajectories of Göreme and Ürgüp at that time was, of course, their relative positioning in relation to the Göreme National Park boundary. Göreme's 'protection' from foreign and other external business investments allowed – perhaps even encouraged – tourism to be developed in entanglement with peasant sociality and limited good beliefs.

This also meant that Göreme as a tourism destination remained remarkably static in terms of the quantity, size, and type of tourism businesses there until the mid-2000s. In my fieldnotes taken during fieldwork in 2005, with only a small number of new businesses having opened during the 1990s and early 2000s and most existing businesses continuing to offer what they had always offered, I commented that Göreme's tourism differed little then from how it had been during the 1980s. This does not mean to say that the community had remained static; for example, by 2005, there had reportedly been as many as 38 marriages between Göremeli men and 'tourist' women, leading to new mobilities and new dynamics by way of connections with elsewhere. In terms of tourism business within Göreme, however, there had only been the occasional 'new' input and idea, such as agencies offering mountain-bike tours or abseiling opportunities in the valleys around Göreme. In 2005, also, a Göremeli entrepreneur who during the 1990s had developed a successful carpet shop which was one of the very few businesses aimed at the tour group market, opened a large 'fine-dining' restaurant on Göreme's main street. As well as serving lunch to the tour groups visiting the entrepreneur's carpet shop, on the weekends particularly, this restaurant attracted 'higher end' international visitors such as international embassy staff based in Ankara as well as wealthy Turkish visitors who were likely staying in Ürgüp and visiting Göreme during the day time. Other than this restaurant and the occasional new activity product, there was little by way of Göremeli *peasant*-entrepreneurs' creating new products or actively seeking new tourist markets.

Indeed, with the image of limited good beliefs being absorbed into tourism, there was neither a vision for expanding into new markets nor a vision of even the possibility of growth; the general approach to tourism business up until the mid-2000s was premised on a logic of subsistence rather than growth and development. This aligns with Van der Ploeg's (2008, p. 117) noting that a peasant mode of production sees 'the market' merely as an outlet, rather than it being the main ordering principle as for the entrepreneurial mode of production. Any growth and business developments that had occurred were developed incrementally and within the business owners', and the community's, means. Importantly, also, with high levels of cooperation and mutual support, there had been little use of credit from external parties to build the early businesses. People such as Abbas and Osman were doing business "within their ability", and younger men, including Mustafa and Mehmet, were still working in others' businesses, albeit busy building social capital if not monetary capital while they did so. Meanwhile, women such as Hanife and Hayriye remained largely occupied with garden work and domestic (re)production, although as told in the portraits above, in 2003, the two sisters set up a small stall next to their valley garden in the hope that making a few modest sales of jewellery to tourists would ease their household's survival. Indeed, premised on an image of limited good, the Göremeli approach to tourism business remained conservative and was focused by and large on reproducing itself; with limited good came limited change. As Bailey put it, peasant societies 'run hard *in order to* stand still' (1971, p. 23). Foster's words, too, once again reflect how Göreme's tourism, during its first two decades, played out in its entanglements with peasant sociality and limited good beliefs

... people in Limited Good societies opt for an egalitarian, shared-poverty, equilibrium, status quo style of life, in which no one can be permitted major progress with respect to any "good." Limited Good behaviour is calculated to maintain the status quo.

(Foster, 1972, p. 58)

Unlimited good and getting beyond the status quo

Although the opening of the first hot-air ballooning company in Göreme in the late 1990s inevitably initiated some changes in Göreme's tourism, it was not until the mid-2000s onwards that a sense of exponential change gained momentum. The growth of the hot-air ballooning sector and the image of unlimited good this growth gave rise to was central to the changes that ensued. By the mid-2000s, ballooning over the Cappadocia landscape had become a major drawcard both for the region and for Turkey as a whole. The front cover of the 2009 edition of the *Lonely Planet Turkey* guidebook displayed a photograph of balloons flying over fairy chimneys in a Göreme valley, and similar images and promotions featured in international newspapers, magazines, and television programmes, including Michael Palin's (of Monty Python fame) *New Europe* television series. By the late 2000s, Cappadocia had become one of the world's major hot-air-ballooning centres, with 11 ballooning companies operating in the region and approximately 40 balloons flying – usually in the Göreme area – each morning. The effects of this ballooning fame on Göreme's tourism was significant, and tourism businesses of all types proliferated. From 15 tour agencies in the village in 2005, the number jumped to 25 in 2009, with most of these vying to sell not only the day tours that they had always sold but ballooning tickets also. There were approximately 65 pansiyons or small hotels and 25 restaurants in Göreme in 2009 and, with accommodation businesses being largely full from spring through to autumn, new ones were opening all the time during this period. Despite the 'global financial crisis' which had occurred in 2008, Göreme's business was booming in 2009, and tourists – largely drawn by hot-air ballooning – were now seemingly plentiful.

Important for tourism growth, also, was that Cappadocia became more accessible during the 2000s; no longer was a twelve-hour bus trip necessary to get there from Istanbul. At approximately 90 minutes' drive from Göreme, the airport in the city of Kayseri, which previously was a military airport, became fully established as a civilian airport by 2009, and had numerous flight connections per day with Istanbul as well as a smaller number of direct flight connections with some northern European cities. In addition, the new and much closer Kapadokya Airport situated near Nevşehir opened, also servicing flights to and from Istanbul. Along with hot-air ballooning fame, the better flight connections to Cappadocia, plus better connectivity via information technology developments, attracted a new type of independent tourists to visit and stay in Göreme. With somewhat bigger budgets compared with their younger 'backpacker' counterparts who previously were the predominant type of tourists staying in Göreme, these more mature tourists preferred more upmarket accommodation and restaurant experiences. To meet their preferences, some higher end but still vernacular in character businesses were developed in the old neighbourhoods, including Mehmet's initial cave-hotel and the smart restaurant attached to it, as well as Mustafa's cave-restaurant which opened in 2011.

At this time, many accommodation establishments which had previously operated as 'pansiyons' for lower-budget backpackers upgraded at least some of their rooms with the aim of attracting these higher-budget guests and also thereby having more certainty of collecting commission from selling them balloon rides. Some accommodation businesses, following the lead of successful entrepreneurs such as Mehmet, underwent complete conversions into "boutique cave-hotels" (see Chapter 5). These conversions involved incorporating ensuite bathrooms into every guestroom, creating a breakfast room with buffet breakfast, and some hotels incorporated a spa, a swimming pool, or a garden into their grounds. While still mainly small in size, these boutique hotels tended to increase in formality compared with the previous pansiyon accommodation. For example, instead of the Göremeli owners running the business themselves as they had previously done, they often employed a manager with hotel management training as well as hiring designated office staff to manage bookings. The new type of independent tourists liked to book flights and accommodation in advance, and they were now able to do so via the websites of the new and converted boutique hotels – which also organised pick-ups from the regional airports to transport guests directly to the hotel. Hoteliers such as Mehmet who employed young, educated people well-versed in digital technology to develop their website and manage the hotel's bookings began to do very well.

As well as the more mature independent tourists from the same countries as the earlier backpackers – predominantly, Western Europe, North America, Australia, and New Zealand – there were increases in other tourist nationality groups, including Turkish tourists who were drawn to Cappadocia during the late 2000s because of a popular television soap opera which was filmed in the area. This upsurge in domestic tourists was also enabled by the relatively stable period of economic growth across Turkey at this time. In addition, the nearby city of Kayseri was experiencing an economic boom as it was becoming a national hub of business and technology, and Göreme and other Cappadocia towns became popular with Kayseri weekend-trippers. Other 'new' tourists included those from various Middle Eastern countries, who – like most of the Turkish visitors – were not especially interested in the dry, dusty landscape of the valleys around Göreme, even if it did include the oddity of fairy chimneys. Preferring greener scenery, after an early morning balloon-ride, these tourists would take what came to be called the 'Green Tours' offered by tour agencies which visited greener valleys in other parts of Cappadocia.

From 2011 onwards, Chinese tourists were identified by entrepreneurs such as Mehmet to be a potential new market for Göreme. It was no longer necessary for tourists from China to visit Turkey on a group tour as they were now able to acquire a visa for independent travel also. Turkey quickly became a popular destination for younger Chinese travellers and, together with a few of his contemporaries, Mehmet was quick to see the opportunity of such new markets. He employed a Chinese student to work in the reception of his hotel as well as manage email bookings in Chinese and to promote the hotel on Chinese social media platforms; these platforms were also beginning to expand in their power of influence. Also influential in promoting Göreme in China was a 2015 Chinese celebrity-travel television show. The programme showed the celebrities staying in a cave-hotel in Göreme and going hot-air ballooning. Later in that year, there was a notable presence in Göreme of tourists from China who were following the itinerary of the celebrity in the programme. Several Chinese restaurants were opened in Göreme at that time also, most of which were at least part-owned by Chinese entrepreneurs and employed Chinese chefs.

Along with Chinese social media platforms, other major platforms – such as *Instagram* – grew in size and in influence during the 2010s. More in-depth discussions follow in Chapters 3 and 4 about Göreme having become a location *Instagram*-users would travel to in order to accrue 'likes' from photographs taken there; the proliferation of social media provided the vehicle for Göreme's significant global fame achievement in relation to the ever-increasing number of hot-air balloons flying over the place each morning. Tourist numbers increased rapidly and business numbers grew to match. By 2014, there were 130 hotels and approximately 45 restaurants in the village, as well as 35 tour agencies, 12 car and motor-scooter rental companies, and 4 horse ranches. The number of tourist shops had also sharply increased, with 16 carpet shops and 40 or so other tourist-oriented souvenir or gift shops. Out of the 21 hot-air ballooning companies operating in the Cappadocia region at that time, eight were based in Göreme itself. Along with social media, digital booking platforms also became fully established at this time. With tourists increasingly using online booking agencies to make their hotel bookings, hotels, along with restaurants and activity businesses were subject to online customer reviews and comments. This in turn altered the relationship between Göreme hosts and their tourist guests – as alluded to by Osman introduced above.

Throughout the 2010s, businesses became increasingly market, or customer, led overall, and consequently, there was a reduction in business owners' sense of control over their business and their tourist guests. This was one of the many ways in which the peasant mode of production weakened during this period. As mentioned above, maintenance of autonomy and control in production are characteristically important in peasant production, and the shift towards more market-led business practices, including the increased formalising of management and running of businesses, meant a lessening of the hands-on owner-operator character of the businesses which had previously been the case. An increased sense of Göremeli people losing control over Göreme's tourism was also associated with an increase in exogenous businesses. The significant capital input required to set up a hot-air ballooning operation meant that many were owned by exogenous entrepreneurs or companies, plus the growth in tourist numbers brought increasing numbers of exogenous entrepreneurs into Göreme more generally. This meant that business competition was no longer predominantly among Göremeli entrepreneurs (who had competed to remain equal), but was now with outsiders' businesses too. In the late 2000s, an Istanbul-based tourism company bought the original Kapadokya Balloons company, as well as one of the very few larger hotels in Göreme catering to bus groups. That exogenous tourism company continued to increase its hold over Göreme's tourism business throughout the following decade.

While tourism became increasingly market-led and the peasant mode of production weakened, many aspects of 'limited good' sociality and behaviour continued – either despite or because of the neoliberalisation occurring. For instance, as the increased presence of 'outsiders' (*yabancı*) in business added to a sense of Göremeli people losing control over tourism, the exogenous-owned businesses simultaneously served to strengthen the felt need for cooperation among Göremeli-owned business owners. Frequently telling me that: "Göreme belongs to us, and we need to keep it that way", Göremeli entrepreneurs continued to group together to form partnerships in order to give themselves stronger leverage in purchasing property and ownership of tourism businesses. This is likely why Osman received financial help from his competing hotelier friend during the 2016–17 tourism downturn period. Other examples of the lingering of 'limited good' sociality and behaviour included

Göreme's more successful businessmen continuing to live an apparently modest life – this aligning with Foster's (1972, p. 58) saying that within 'limited good' sociality, there must be 'concealment or denial of "good" possessed' because to incite jealousy would incite criticism or even ostracism. Similarly, whilst appearing to be partly said in jest, Mustafa's expressing discomfort about his restaurant being No.1 on *Tripadvisor* could have been an expression of his discomfort at getting beyond approved levels of mediocrity; if success cannot be concealed or denied, then it should at least be downplayed in order to give the *appearance* of mediocrity, as illustrated by the wealthier Göremeli businessmen continuing to drive around the village on a small moped. 'Limited good' sociality thus lingered on within Göreme's tourism business, despite its having less of a firm hold as tourism 'good' increasingly appeared as *un*limited.

Moreover, the growth in tourism created some tangible reminders of actual limitations to 'good'. As more and more houses in the old neighbourhoods were being converted into hotels, for example, the limits to available Göreme land and the imposition of National Park regulations became increasingly apparent; unable to buy land to build a new house in Göreme, Göremeli families departing from the old neighbourhoods became displaced to other towns and villages. The growth of tourism also highlighted water as a limited good. Frequent water-shortage problems in Göreme affected residents in their homes as well as the hotels and restaurants. Indeed, the proliferation of boutique hotels developed with gardens, swimming pools, and guestrooms with attached 'luxury' bathrooms has confirmed that unfettered growth will, in the end, create problems for *everyone*. Throughout the 2010s, Göremeli businessmen frequently came together to discuss the water-shortage problems and how they might be remedied. During such discussions, they voiced suspicion that the hotels owned by non-Göremeli entrepreneurs and companies had widened their pipes to capture a greater volume of the water coming into Göreme. It was also suspected that an increase in potato farming on the plateau where Göreme's water supply originated meant that the only solution might be for the Göreme Tourism Development Cooperative to buy up the potato fields so that the land could be returned to water catchment for Göreme. In the meantime, water that trickled underneath Abbas's house was pumped out and taken by Emre to hotels situated higher up in the village, such as Osman's hotel, where it was needed to keep the garden watered and for guests to fill their ensuite jacuzzi-bath. Hence, 'limited good' principals have been fully borne out in relation to water.

The above examples highlight how peasant production and sociality on one hand and global tourism capitalist processes on the other became entangled in such a way as to create tensions and ambivalences in each other. The Tourism Development Cooperative's becoming increasingly business-oriented – developing a *hamam* (Turkish bath) and shop complex in the village centre – is a further example. So too is Hayriye and Hanife's stall and shop business development which, although occurring a decade or two after the – predominantly men's – peasant-business developments referred to in the 'tourism as limited good' sections above, was intended – initially at least – merely to ease their household's survival. Moreover, the two sisters' simultaneous running of a valley stall and a central Göreme shop in order that they could 'catch' tourist customers in both areas – the valley and Göreme's main street, aligned with the limited good principle of 'catching' as many of the limited available tourists as possible. It was not until several years after opening the initial business that they began to see a possibility of finding, even creating, new markets for their jewellery products, which they did by securing commissions with individuals and companies overseas. Rather than seeing 'the market' merely as an outlet as per the peasant mode of production (Van der Ploeg, 2008, p. 117), Göreme's entrepreneurs were slowly becoming more market-led, both in the sense of 'the market' becoming an ordering principle for their business (as Hayriye and Hanife took on more commissions, they needed to recruit more women in other villages to raise production levels to meet demand), and in them beginning to view the market as extendable. This marked a significant shift in Göreme's peasant-tourism-economy in the period post circa 2005, when possibilities for creating wealth appeared not only to be extend*ing* but also extend*able*.

'Depeasantisation' was occurring more broadly across Turkey at this time so that, beyond tourism, there were other major shifts and reforms that contributed to the substantial social change occurring in Göreme. For instance, Turkey's adoption of European Union accession criteria during the early 2000s resulted in significant gender equality reforms, including remarkable gains in access to education for girls and young women (Durakbaşa and Karapehlivan, 2018). While these education reforms came too late to affect Hanife and Hayriye's schooling, the broader changes occurring in Turkish society would undoubtedly have influenced the two sisters' ability – and confidence – to open their own shop on Göreme's main street (see Chapter 6 for further discussion on this). Other development policies since the millennium included an intensive effort in Turkey to improve and develop information and communication technology infrastructure, with availability of mobile phones increasing to 98.7% in 2018 nation-wide while, by 2019, households with internet access increased to 88.3% nation-wide, including in rural areas (Hovardaoğlu and Çalısır-Hovardaoğlu, 2021). Hovardaoğlu and Çalısır-Hovardaoğlu (2021) also cite major agricultural reforms in Turkey during this period which enacted radical changes in agricultural production structures to support a transition from subsistence farming to 'professional' agricultural production (ibid., p. 90). It is inevitable that such national-level reforms would have significantly contributed to social change, with much of that change markedly improving the lives of a great many people, especially the lives of girls and women. In later chapters – Chapter 6 in particular – I discuss how such change has manifested for girls and women in Göreme.

The suggestion by Nash (2007) that much of the world's peasant populations 'are now facing elimination in the global markets of the third millennium' (p. 4) would thus very much appear to apply to Göreme's (former) peasant population. Nash's warning that 'the spectre of the unlimited growth cultivated by neoliberal market economics benefits only a few' (2007, p. 4) also appears apposite. During the 2000s and early 2010s, competition between those in the position to receive commission from ballooning became increasingly fierce and individualistic, while businesses became increasingly growth and sales-oriented. Moreover, with no major global or regional 'crises' affecting tourism in Turkey from the late 2000s through to 2015, plus Göreme's hot-air ballooning fame by then fully established, a perception of tourism and tourists as *un*limited was taking hold. Increasing numbers of people obtained bank credit in order to extend their business, or business portfolio, and the gap grew between those who owned business and property and those who did not – as did the gap between success and non-success in tourism business more generally. Contrary to the level playing field which had characterised the earlier years of Göreme's tourism, significant inequalities emerged with some businesses and individuals getting markedly ahead of others. Some of Göreme's more successful business owners, being able to constantly expand their ownership and wealth, grew very wealthy indeed; Mehmet who was introduced above could be included in this group. Conversely, with wages low and little or no job security, households or families who did not have property or ownership of a business often struggled to make ends meet. This group included increasing numbers of people from other Cappadocia towns and villages, as well as (mainly Afghani) refugees, who sought work and tried to eek a living in Göreme's tourism. Overall, in contrast to the image of limited good which 'balances the collective needs of a population with available land and resources' (Nash, 2007, p. 4), the increased spectre of unlimited good in Göreme's tourism and an associated expanded vision of growth were ultimately leading to winners and losers within the Göreme community.

Some challenges to tourism business emerged once again in the mid-2010s, however. In 2015, the Syrian crisis led to a massive influx of refugees coming over the border into Turkey, and while this mostly did not affect tourism places and provisioning in any direct way, it kept many European and North American tourists away. Things worsened further in early 2016 when a series of bombings occurred across Turkey, some of which were targeted at tourists, including a bomb attack in the Sultanahmet district of Istanbul which killed several German tourists, and a bombing at Istanbul's Ataturk Airport in June of that year. Then, in July 2016, tourism came to a sudden halt with the attempted coup in Turkey. When I visited Göreme in October of that year, the streets were empty of international tourists, and there were only a few Turkish tourists travelling around. With Turkey considered still too dangerous to visit, 2017 continued in the same way. Throughout 2017, Chinese citizens were told by their own government not to go to Turkey because of the perceived danger.

During this time, Göreme hoteliers and other business owners were grateful that a trickle of domestic tourists kept most businesses afloat. At the start of 2018, the continuation of the same minimal level of business was all anybody dared expect. That apprehension was likely what led to two-thirds of Cappadocia's ballooning operations entering into an arrangement whereby their flight operations leased to the Istanbul-based tourism company (the company mentioned above which had previously purchased Göreme's original hot-air ballooning company). Contracted to last for three years, this arrangement appeared to offer a lifeline to the ballooning operations at a time when tourism business was extremely slow. However, as the year 2018 progressed, things turned out to be very different indeed. A previously planned for 2018 China-Turkey 'Year of Tourism' partnership which had looked unlikely to go ahead did happen after all so that Turkey went from being prohibited for Chinese tourists in 2017 to being strongly promoted in China in 2018, prompting a significant influx in that year of Chinese tourists to Cappadocia.

Additionally, the drop in tourist numbers during 2016–17 had prompted Mehmet to ramp up his efforts to boost Göreme's presence on social media. Along with his actively working *Instagram* and other platforms as well as hosting social media 'influencers' at his hotels, Mehmet was proactive in extending Göreme's fame into the Russian tourist market, inviting groups of models from Russia who would then act as social media 'influencers' in that market. Moreover, Mehmet's *Instagram* and other social media work ensured that, among all of the towns in Cappadocia, Göreme became known as the best place to stay in order to be able to get up in the early morning and, from your hotel's breakfast terrace, watch – and create your 'likable' social media postings among – the one hundred balloons flying overhead. In 2018–19, with tourists clamouring to stay in Göreme, previous ideas of tourists and tourism as *limited good* fully gave way to an image of tourists as *more than plentiful*.

The more than plentiful tourists meant that, along with Göremeli vying to increase their business foothold, increasing numbers of people from elsewhere were drawn to Göreme either to do business or to gain employment in tourism. While many incomers bussed into work in Göreme every day from nearby towns and villages, and some workers had come from as far as Afghanistan, a few incomers were in another league altogether. Significant among these incoming large players, albeit 'incoming' in a virtual way, were the major online booking platforms such as *booking.com* and *Expedia*, which most Göreme hotels had come to use. Other – more tangible – examples of incoming large-player business interests included the Istanbul-based tourism company mentioned earlier which, having secured a three-year lease arrangement on the majority of the region's hot-air balloons, began also to buy hotels in Göreme. Additionally, in 2019, an exogenous company which was rumoured to have ties to the governing Justice and Development Party (*AKP*) purchased a collection of old houses adjacent to Göreme's main street and constructed a large

Figure 2.5 New hotel deemed to be "spoiling the view" on Göreme's main street.

high-end hotel there, directly opposite the old Göreme teahouse where mostly retired and elderly men spent their days playing backgammon. While the men bemoaned the fact that the new hotel was too big and hence was out of place in Göreme, completely "spoiling the view", they also knew that it was unlikely that anyone would listen to their complaints. Earlier that year, the ruling *AKP* had established a new central governance body, the 'Cappadocia Area Directorate', to oversee preservation zoning and building permits in the Cappadocia region. With this move removing planning and building decision-making powers from Göreme's Municipality Office, suspicions that the new directorate's main purpose was to protect the business interests of external investors did not seem misplaced; later in 2019, Göreme National Park was stripped of its national park status by Presidential decree. The Göreme National Park's disestablishment would constitute a pivotal move in terms of removing any remaining limitations on Göreme's tourism; hence, tourism had become for all concerned – Göremeli and non-Göremeli – an unlimited good.

Jonah, excess, and the spectre of unlimited good

Jonah: 'a tale of ambition and arrogance in the world of international tourism. Mbwana and Juma have dreams for their underdeveloped home town, which isn't the tourist destination they'd like it to be. Suddenly, they strike gold – or, rather, fish. An enormous sea monster jumps from the water as they walk along the beach and they miraculously capture it on camera.

Such a discovery turns out to be a perfect gimmick for bringing in tourist dollars, and Mbwana leads the charge. Yet, as seems to often happen…, the moral quickly becomes "be careful what you wish for." Director **Kibwe Tavares** *uses the lushest of effects…, morphing the city into a chaotic, seething wall of neon advertisements".*[9]

Just before visiting Göreme in October 2018, I had been introduced to the short film *Jonah*. The film tells the story of how the dreams of Mbwana and his friend Juma become reality when they photograph a gigantic fish leaping out of the sea and, consequently, their small town morphs into a decadent tourist hotspot. However, with its sleazy bars and giant neon signs reading "Big Fish Mall", "Fish House Bar", and "Fishbucks Coffee", the previously quiet town, and its community, soon deteriorate into seedy oblivion. When I arrived in Göreme a few days after seeing the film, the evening darkness was starting to set in as my airport transfer wove around the Göreme streets dropping off tourists at their various hotels. It had been two years since I had last visited Göreme and I was shocked by the extent of the chaotic gaudiness. Neon signs in English, Turkish, and Chinese characters advertised balloon rides, ATV tours, noodle bars, and Chinese, Korean, and Indian food. The resemblance of Göreme to the town's neon excess in *Jonah* was striking.

Indeed, 'excess' seems apposite here precisely because of its conjuring the notion of something being beyond limits, and beyond 'enough'. Tourism in Göreme had more than recovered from the one-and-a-half year hiatus that had followed the 2016 attempted coup, and just as tourist numbers multiplied once again, so did the number of tourism businesses. In the years 2018–19, the number of hotels in Göreme reached over 200, while cafes and restaurants numbered more than 100 – including 13 Chinese restaurants and 4 large restaurants on the main street with live music licences. In addition, there were 85 tour and transport booking agencies, 20 carpet shops, 60–70 other tourist souvenir shops, 10 rental car shops, 8 balloon companies (in Göreme itself, while 20 more companies were based in nearby towns), and 5 horse ranches. No longer was tourism business in Göreme maintained at an approved level of mediocrity. As Abbas and his friends had rightly remarked, no longer could Göreme be regarded as a *köy* (village), and nor could its population be regarded as *köylü* (villager/peasant). This is further illustrated in the following extract from my 2019 fieldnote diary.

Fieldnote diary, 2019: Yesterday evening at around 7 pm as I was walking back to Aydınlı Mahallesi and passing through the bus station area, there was a scene of utter chaos. Gridlocked buses and cars, nose-to-nose with each other so that no one could move at all, horns beeping, a döner-kebab vendor calling out to the scores of tourists waiting to board overnight buses headed for Istanbul or Antalya, an ice-cream seller's shouts of 'Ice-cream – nice-creeeam!' were reminiscent of the area in front of Hagia Sophia in Istanbul. I stood and watched for a few minutes, marvelling at the mayhem – and at the ridiculousness of it all.

Continuing on up towards Aydınlı Mahallesi, as I walked up the narrow street entering the old neighbourhood, that too was gridlocked with horn-beeping traffic and tourists heading out for dinner trying to navigate a safe path for themselves amongst the jammed up cars and airport shuttle-minibuses. Filiz, who sells jewellery at her doorstep, was standing watching the chaos and, stopping beside her to watch too, I said – having to shout above the racket of beeping horns – "Istanbul gibi!" (It's like Istanbul!). She replied "Evet, Göreme kent oldu" (Yes, Göreme's become a city"); "Look at all the expensive cars – they are all expensive cars", Filiz continued, "It's easy for people to get rich here. They got rich quickly, and easily!"

In telling me how exhausted she was, all of a sudden Filiz burst into tears, seemingly beside herself, distraught not only at what had become of the place, but also at the sheer unfairness of it all: "The hotel next door, they just sit and do nothing all day and earn a trillion lira – from doing nothing!", she said, "They get rich too easily and don't have to lift a finger; It's not fair!" She then explained that she was not saying this out of jealousy, but rather because "it's not right": "They don't have to lift a finger, whereas I can only afford to employ one Afghan worker, not even two, just one, so I have to work alongside him clearing the debris out of my house to renovate it"; "I felt sorry for him, sweating and carrying rocks all day – for such little money. The Afghans are the ones doing all the hard work here now – can you imagine Göremeli men working hard like that these days? No! None of them would do it now, because they've got it too easy now."

It has been argued in the recently emerging literature on 'overtourism' that the phenomenon – where locals or tourists feel a deterioration of place due to there being too many tourists – occurs 'as a result of uneven accumulation of and concentration of resources and space... and is very often embodied in the reproduction of inequalities' (Milano, Cheer and Novelli, 2019, p. 8). As already mentioned in Chapter 1, in these respects the term appears apt in relation to Göreme. Within just a decade, Göreme had become a bustling town filled with the comings and goings of social media following tourists and people from elsewhere drawn to the town in order to profit from tourism. Yet, the fieldnote diary extract above, and especially the plight of Filiz, is a reminder of the importance of considering the multiple, and uneven, entanglements through which tourism becomes a part of people's lives. Indeed, the 2010s decade had seen something of a race whereby, with tourist numbers seemingly without limits, both exogenous *and local* interests clamoured to buy, expand, or open new business. The notion of 'overtourism' does not seem even to begin to capture these ambiguously entangled dynamics in Göreme's tourism, nor the ambivalent ontological tensions which inevitably come with the prospect of unlimited good, and unlimited change.

To refer once again to the words of Nash which opened this chapter, 'the notion of the unlimited good, while raising the levels of expectation for a better life, may also cultivate behaviour that raises a specter [*sic*] of doom for many' (2007, p. 41). Whilst in the late 2000s only a very small minority of Göremeli entrepreneurs could have been considered wealthy, during the following decade many more locally owned businesses and entrepreneurs succeeded in getting beyond the level of mediocrity which had previously been sustained as a steady and slow-burning equilibrium. Consequently, Göreme became a place of contrasts between winners and losers, haves and have-nots, wealth and precarity. Younger entrepreneurs who became tech-savvy were able to work social media in order to extend their market share and thereby boost their business. The temptation of many to go beyond their means in order to compete in the 'business-race' was no longer able to be resisted, and many people took bank credit and other loans in order to expand their business endeavours. With tourists having become *more than plentiful,* for those who dared, being in debt to an external organisation was now considered worth the risk.

It is difficult to say whether such Göremeli business owners could still, after four decades of tourism in Göreme, be considered even 'semi-peasant'. However, even as processes of depeasantisation and a shift towards a capitalist market economy were increasingly embraced, elements of peasant production and sociality do appear to still persist; aligning, perhaps, with Öztürk et al.'s (2018) assertion of Turkey's 'new peasantry' being a 'non-homogenous category'. Some families, especially and perhaps paradoxically those who had become wealthy from tourism, have kept hold of their gardens and continue to grow vegetable and fruit crops; hence, traces of a Göremeli 'internal communing ethos' (ibid., p. 251) linger on. Moreover, with a high percentage – estimated at 80–90% – of tourism businesses still in local ownership, remnants of that internal communing ethos and peasant moral economy continue to create fissures in tourism's global capitalist processes. For example, a seemingly deep-seated drive to give hospitality to tourists can still – frequently – evade the drive for profit-making, and neighbouring hoteliers still lend money to each other during tough times, likely in recognition that next time it may be themselves who would need support to keep afloat. Meanwhile, those Göremeli families who have not become wealthy maintain that they nevertheless have "enough". All in all, some traces of peasant sociality and forms of production appear to continue – apparently both because of and despite global capitalism in the form of tourism.

As the 2010s decade had proceeded, however, an increasing sense of such continuities collapsing under the weight of change had developed. Tourism business began to fill not only Göreme's central streets but the old neighbourhoods also, causing the value of land and property – especially in those old neighbourhoods – to rise exponentially, resulting in displacement of residents through processes of gentrification (see Chapter 5). As well as houses being sold for conversion into hotels, parcels of land situated all around the edges of Göreme – many of which were orchards and vineyards that had been handed down through generations – were sold to hot-air balloon companies. In the interests of tourism capital, those gardens were stripped of their decades-old grapevines and apricot and walnut trees to become empty spaces for use as take-off sites for balloons. Perhaps the matter of whether Abbas's return to farming means he has become once again a 'peasant' is, after all, irrelevant when he tells me that the leaves on his grapevines cannot breathe because of all the dust coming from the take-off sites that now surround his primary vineyard. When it is not the dozens of vehicles coming every morning in association with ballooning creating the dust-clouds, it is the scores of tourists on quadbikes (ATVs) using the stripped swathes of land for their fun every afternoon. To refer back to Abbas's analogy referred to at the start of this chapter, whereas the "new path" he and his friends had cut was a path of limited good, the "huge highway" it became was both a manifestation of unlimited good, and an emergent foreboding spectre of unlimited change.

Notes

1 In *Living with Tourism (1)*, I gave Osman the pseudonym Ömer (see pp. 93–94), while for this book, he would prefer I use his real name.
2 In separate parts of *Living with Tourism (1)*, I gave him different pseudonyms – Ali (see pp. 71–72) and Hüseyin (see p. 95). In this book, he would prefer I use his actual name.
3 For as long as I have been studying Göreme, its population has hovered around the 2000 mark, which in Turkey is the population required for a village (*köy*) to become a *kasaba* and thereby able to have its own *belediye*/municipality office and hence a certain level of autonomous governance.
4 See section on 'Crisis of *our town*' in Chapter 8 for related discussion.
5 Öztürk *et al.* (2018) clarify these translation points in the following: 'Although there was/is no word for 'peasant' in the Ottoman language or modern Turkish – the nearest being 'rençber' (poor farmer, rural labourer) and 'köylü' (villager) – this term can be applied to the Anatolian situation in its generalised meaning of smallholder/subsistence farmer, and 'peasantry' is a reasonable translation of 'köylülük" (p. 244).
6 Tosun (1998) had discussed how the growing predominance of larger, "all-inclusive" hotels in nearby Ürgüp during the 1980s–90s forced many of the smaller, locally owned tourism businesses there to close. Similar patterns reportedly occurred in other tourism destinations around the world, including Pattaya in Thailand (Wahnschafft, 1982), Ladakh in India (Michaud, 1991), and Bali in Indonesia (Picard, 1996) as well as several locations throughout the Mediterranean region, including on the Turkish coast (Egresi, 2016).

7 The activities and shopping locations of tourists on group tours tended to be tightly controlled in order that the tour company and/or the guide would receive a hefty commission in relation to whatever they did or bought.

8 While an early discussion of 'touting' was provided by Crick (1994) in his descriptions of street-guide touting activities in Sri Lanka, it is surprising that such practices have so seldom been discussed in the tourism literature. Most discussions of commission practices in tourism appear to center around (Chinese) tour group practices rather than touting per se (Ap and Wong, 2001; King, Dwyer and Prideaux, 2006; March, 2008).

9 This extract is from a description of the film *Jonah* on short-film review website: https://filmschoolrejects.com/short-film-jonah-is-a-chaotic-gorgeous-fish-story-fd6aa84e2492/

3 Ups and downs of hot-air balloons over two decades

> Tourismscapes are materially heterogenous and consist of bounteous people and things interacting or networking…, and concurrently enacting tourism.
>
> (Jóhannesson, Ren and van der Duim, 2012, p. 166)

The apparent resemblance of Göreme to the Tanzanian town's tourism trajectory in *Jonah* (introduced in the final section of Chapter 2) led me to watch the short film again during my 2018 visit to Göreme, this time with the recently retired Göreme businessman Mehmet.[1] The film appeared to render him somewhat stunned. He said, 'Mmmm… it's like Göreme's story, except here instead of the big fish it's the balloons'. Mehmet's words raise some important questions: Was the ultimate demise of the town in the film caused by tourism? Or was globalised capitalism primarily to blame? Or was it 'Fish Man' Mbwana? Could the gigantic fish itself be the culprit? Can a fish be considered a culprit? If so, what does it mean when Mehmet says, "it's like Göreme's story, except here instead of the big fish it's the balloons"?

Of course, the tourism trajectory of the town in *Jonah* can be put down to, what Franklin (2012) might say is, a combined choreography of a great variety of 'dancers'. Along with 'Fish Man' Mbwana, a host of business-interested parties, plus globalised capitalism, tourism, mass consumerism – and of course the big fish – can *all* be counted among the bounteous people and things interacting or networking to concurrently enact tourism (Jóhannesson, Ren and van der Duim, 2012, p. 166). Nonetheless, Mehmet's comment was interesting, as was a similar idea conveyed by Mustafa, the director of the Göreme Tourism Development Cooperative referred to at the start of Chapter 1, who said: "It's the balloons which are the problem. They are such big money, that's why everyone's turning to Göreme to do business – including big companies, it's all to do with the balloons". The words of Mustafa and Mehmet beg the question: While undoubtedly there are bounteous people and things concurrently enacting tourism, in what way, exactly, might the balloons themselves be seen as the central protagonist enacting tourism in Göreme?

DOI: 10.4324/9781003011200-3

Actor-Network Theory (ANT) could provide a useful perspective from which to approach such questions, since "ANT does not see the human actor as the sole or primary 'mover and shaker'" (Ren, Jóhannesson and van der Duim, 2012, p. 16); rather, ANT sees material entities and non-human animals as having 'the capacity to act just as well as humans' (ibid., p. 16). Therefore, bringing ANT's relational ontology into tourism contexts means, as Simoni (2012) puts it, that 'bodies, objects, spaces, and media technologies are accounted for... as agents that enable and actively constitute tourism, informing the way "tourism ordering" of the world operates' (p. 61). Hence, a variety of non-human protagonists have been discussed in relation to the ways in which they have informed and actively constituted 'tourism ordering'. These include coral reefs (Middelveld, van der Duim and Lie, 2015), gorillas (van der Duim, Ampumuza and Ahebwa, 2014), cigars (Simoni, 2012), cheese (Ren, 2011), floating markets (Pongajarn, van der Duim and Peters, 2018), buckets and spades (Franklin, 2014), backpacks (Walsh and Tucker, 2010), and headlice (Benali and Ren, 2019). In this chapter, I will consider the way in which the hot-air balloons have – as primary 'movers and shakers' – enacted tourism in Göreme over the space of two decades.

To do so, in alignment with ANT's emphasising of multiplicity and complexity, also, rather than considering the balloon-protagonist as singular, I will draw upon ANT's method of accounting for the ways in which different versions or configurations of 'things' (and non-human animals) enact multiple relational networks. Previous examples of this include the depiction by van der Duim, Ampumuza, and Ahebwa (2014) of five configurations of the gorilla network in which gorillas are designated as 'trophies' in the hunting network, as 'man's closest neighbour' in the scientific network, as 'endangered species' in the conservation network, as coexisting with local livelihood practices, and, finally, as part of the tourism network. Similarly, Ren (2011) discussed how, in an area of Poland, a variety of versions of osckypek cheese – 'traditional', 'tourism', 'modern', and 'unique' – enacted sometimes aligned and sometimes contested orderings. In relation to Göreme's tourism, I have identified five key hot-air balloon configurations – '*Elite balloons*', '*Money-maker balloons*', '(*Un*) *regulated balloons*', '*Nuisance balloons*', and '*Instagram balloons*' – which make up the multiple entangled 'tourism orderings' in Göreme as constituted by the balloons. Each configuration, or balloon 'character', conveys a different way in which the balloons have enacted Göreme's story, or rather stories, over two decades.

Moreover, along with recognising the simultaneity of multiple balloon-enacted 'tourism orderings' in Göreme, since I am taking a longitudinal perspective in this book, it is necessary to include the important dimension – or actor – of time, and thereby to consider how the various tourism orderings play out over, and perhaps under, time. This brings to mind the image of woven fabric, whereby each balloon 'character' is like a thread which weaves in and out, up and down through the, also interwoven, years. Indeed, Ingold (2011) suggests 'meshwork' to be a more apt translation of *réseau* in Latour's (1999) *acteur réseau* (actor network): '*réseau* can refer just as well to netting as to network – to woven fabric, the tracery of lace' (Ingold, 2011, p. 85). In this chapter, I assemble the different balloon characters' 'movings and shakings' within three main 'time', or decadal, sections – circa 1999, 2009, and 2019 – so that, within each decadal section I discuss the various balloon configurations that were most 'active' at that time. In this way, I depict what I see as a kind of 'balloon-time meshwork', with different balloon characters coming into being at different times, and some then receding or dissipating altogether, in an 'ever-raveling and unravelling relational meshwork' (Ingold, 2011, p. 142).

Circa 1999

1999: Elite balloons

> Extract from article on hot-air ballooning in *Forbes Life Magazine*, May 1998: The good landings, which occur more than 99.9% of the time, usually go something like this: your chase crew have successfully followed you in their van to the spot where you want to touch down. You toss them your rope, they brake the balloon's vertical motion and you bring the balloon down as gently as a glass of water. The locals stream out of their houses, offices and fields, and in their enthusiasm start setting up an impromptu feast of the best indigenous booze and goodies they can lay their hands on. You, the pilot, step nimbly from the basket, and are acclaimed by all as some sort of divine hero from the skies; Apollo descended from the heavens.
>
> –
>
> Fieldnote diary extract, summer of 1996: I was woken very early this morning by a strange noise outside. I looked out of the window and a huge balloon filled my vision. It had landed on the road next to the house I was staying in and was causing quite a commotion, not due to any problem such as blocking traffic, but rather because of its sheer size and magnificence. People came out of the hotel next door, where Kaili and Lars – the balloonists – were based, and made a fuss of the wondrous big thing, while the faces of the tourist passengers in the basket beamed with pride. After a while, the pilot, Lars, pulled the gas lever igniting the burner (making me realise what the strange noise was that had awoken me) and the balloon and its passengers lifted off the ground, majestically rising into the air – gigantic and yet graceful, and now silent, it slowly drifted off out of view.

Along with their, at first one and later two, balloons, Kaili and Lars were key actors in the early years of the balloon-time meshwork. From northern Europe and both experienced hot-air balloon pilots, the couple set up the first ballooning business in Göreme in the mid-1990s. Initially they operated out of one of the larger hotels situated just outside of the Göreme National Park boundary, and soon afterwards, they moved into the village itself and operated from Göreme's – only at that time – boutique hotel. In 1997, with the support and partnership of some non-Göremeli Turks who helped them navigate the Turkish business set-up bureacracy, Kaili and Lars established their company, Kapadokya Balloons. The company's office and reception was situated adjacent to the main street at the entrance of Göreme, and the site also included a balloon maintenance workshop plus parking areas for the overland vehicles and trailers that carried the balloons and baskets. With their home situated upstairs above the office, the area at the front of the office was also Kaili and Lars' garden, where most afternoons they would sit around a large wooden table and chat with friends, business associates, customers, and other 'elite' visitors. Highly visible to those arriving into Göreme on the main road from Nevşehir, the site became something of a 'balloon hub', which – during the late 1990s and early 2000s – became the place to be and to be seen.

Whilst Kaili and Lars were down-to-earth characters themselves, the Kapadokya Balloons hub had frequent elite business visitors, such as Germans visiting on behalf of Mercedes Benz (from where the four-wheel drive, off-road capable vehicles used to tow the balloon trailers were mostly purchased), or national and international media photographers, journalists, or celebrities who had been invited for promotional reasons. Diplomats visiting from Ankara or members of the Turkish political elite also came by fairly regularly; having flown with Kaili and Lars in the morning, they would pop by in the late afternoon for a gin and tonic, sitting around the wooden table, next to which were the trailers holding the *Elite balloons* wrapped up ready for the next morning's flight. In September 1997, Turkey hosted the Federation Aeronautique Internationale (FAI) '1st World Air Games' and the hot-air balloon section was held in Cappadocia. Sixty-two balloons participated and, even though Kaili and Lars did not take part, throughout the week of the games, the wooden table became a meeting point of the international balloonist elite. Occasionally, Buddy Bombard – an internationally renowned balloonist and owner of a luxury ballooning holiday company – would visit. To explain the level of elite status Buddy Bombard held, the 1998 Forbes Life magazine article quoted above wrote that, if ballooning were a religion, "Buddy Bombard would rate at least a Pope"; his visits to the Kapadokya Balloons hub always created a buzz of excitement.

Göre villagers tended not to be so present during the afternoon drinks hour; a table at which politicians or business elite sat was not a comfortable place to be for most villagers. That said, many villagers were proud to be considered friends of Kaili and Lars and would drop by at other times of day. Moreover, Kapadokya Balloons was considered a good employer of Göremeli people, being one of only a few tourism employers in the village at that time who included full social-security and pension payments. Several men were employed

Figure 3.1 Elite balloon flying over fairy chimneys and vineyards near Göreme, 1997.

full-time as drivers and ground crew for the ballooning operation, plus a handful of mini-bus owners were contracted to ferry tourists around before and after flights. Although working with the balloons necessitated a shift to one's daily timetable and curtailed the social life of younger crew members, being an employee of Kapadokya Balloons held considerable prestige. A couple of the crew members were proud to also be given pilot training by Kaili and Lars, and so began the transition of local men into the elite occupation of pilot.

Indeed, flying in a balloon was an elite activity for all involved and, with the cost of a flight with Kapadokya Balloons being US$250, particularly so for the tourists who took a flight. In the late 1990s, Göreme's tourism still largely attracted a backpacker market, and so it was only a small elite segment of tourists who could, or would, buy a balloon flight while visiting Cappadocia at that time. Those who did were afforded a particularly exclusive view of the Cappadocia landscape; it is a longstanding trope that an elevated gaze is a way of achieving domination and mastery over a landscape (Pratt, 1992; Smith, 2021; Urry, 2002). Moreover, with the idea that 'by seeking distance a proper 'view' is gained' (Urry, 2002, p. 147), there was a perception that the higher up the balloon went, the bigger and better the view of Cappadocia achieved. Kaili and Lars were masterful at working with both the air currents and the large mass of the balloon envelope to choreograph each flight to achieve a variety of heights and angles over different valleys and villages. While the flight path of the balloon would depend largely on the direction of the air currents, by opening and closing different air-vents in the balloon, they could rotate the basket in order to allow those on each side an opportunity to photograph the best views. Through comments made about "how lucky we all are today", the pilots would always ensure that the passengers knew they were having one of the best flights of the season. Then, wherever they landed – usually in an orchard or vineyard – an immediate celebration was held, with cake, champagne corks popping, and certificates handed out, all ensuring that the air of prestige was maintained for as long as possible. That air would continue on after the passengers were dropped off at their hotel where, seen arriving in a Kapadokya Balloons-marked jeep, they would be greeted with envious questioning about how it was by both the hotel's staff and their fellow hotel guests.

1999: Money-maker balloons

As well as the Kapadokya Balloons owners, employees and contracted mini-bus drivers directly earning their livelihoods from ballooning-related activities, the balloons quickly became money-makers for many others in Göreme through commission payments. As already mentioned in Chapter 2, during the 1990s, a complex commission system developed among Göreme's tourism businesses which involved passing on customers to other businesses in return for commission payments. This arrangement mostly occurred with guests of pansiyons and hotels being "sent" on a day-tour with a particular tour agency, or being taken to a particular carpet shop.

The early years of the ballooning commission practice manifested similarly so that, while it was predominantly pansiyon or tour agency owners "sending" customers to Kapadokya Balloons, occasionally tourists would be guided to the balloon office by a waiter or someone they had passed in the street and stopped to ask for directions. Hence, anyone who was in a situation to be able to send or physically take tourists to buy balloon-flight tickets had the opportunity to make money from the balloons; with ballooning tickets priced at US$250 per person, the percentage paid in commission to anyone sending or recommending tourists was a sizeable earning for most. Moreover, at the end of the 1990s, a small number of other balloon companies were being set up in nearby towns and so, although Kapadokya Balloons was still the only ballooning operation in Göreme itself, commission payments became an increasingly important strategy in competing for customers. People regularly dropped by the office to pick up their "envelope" containing a commission payment, and every now and then, confusion and conflict would occur when two or more parties tried to claim commission for the same tourists.

As the balloons came to afford a wide variety of tourism workers and business owners the opportunity for making some extra money, as money-makers, the balloons also came to incite jealousy. Conflicts sometimes occurred among businesses or individuals who were more, or less, capable of selling balloon tickets, and Kaili and Lars themselves also incited considerable jealousy as *Göremeli* people calculated their daily takings in dollars. The anecdote told in the last chapter about business jealousy displayed by the owner of a hotel situated next door to the Kapadokya Balloon office whereby he refused to recommend his hotel guests to go ballooning illustrated how Kapadokya Balloons became mixed up in the competing-to-remain-equal business sociality discussed in Chapter 2.

It was not only Kapadokya Balloons but, we could say, the *money-maker balloons* themselves that became embroiled in the 'image of limited good' beliefs that were very much entangled within Göreme's tourism at that time. In other words, along with a number of the predominantly backpacker tourists – and their individual budgets – appearing to be limited, in the late 1990s, when hot-air balloon tourism in Göreme was in its early stage, the extent to which the balloons could act as *money-makers* – more broadly than the Kapadokya Balloons company and its workers – was viewed as limited too.

1999: (Un)regulated balloons

When Kapadokya Balloons started up, it inevitably incurred regulation and was required to register with the Turkish Civil Aviation Authority in a similar way as any 'airline' would. Nonetheless, compared with the regulatory environment in the parts of northwest Europe where Kaili and Lars trained and had undertaken most of their ballooning previously, flying in Cappadocia was relatively unregulated. With no established hot-air ballooning industry in Turkey at that time, the Kapadokya Balloons pilots worked with the Turkish Civil Aviation Authority to advise and create the level and type of regulations they deemed necessary for theirs and similar businesses. This largely determined the way in which ballooning in the region would develop. For instance, for many years, it remained at the pilots' discretion as to the suitability of weather conditions for flying each morning, as well as where and when they flew. Each morning, Lars would let a small helium-filled balloon drift up into the air and watch it so that he could determine the wind conditions at different altitudes and then decide on the take-off site and flight plan for that morning. There were very few mornings throughout the season in which the conditions were deemed unsuitable for flying, since stable, continental climatic conditions prevail in Cappadocia.

Additionally, since there were only a small handful of balloons flying in Cappadocia at any one time during the late 1990s, many aspects of the flights, such as take-off sites, landing sites, and flight plans, remained unrestricted. The *(un)regulated balloons* of Kapadokya Balloons could take off pretty much anywhere and they often used the bus park areas of the two nearby World Heritage Site museums (Göreme and Zelve), or empty fields on ridge-tops. Being recognised by the authorities that a balloon's flight path was very much at the whim of the direction and speed of the wind each morning, it was accepted that landing could occur anywhere. For the pilots, the most important thing was to land in a place accessible to the four-wheel drive vehicles needed to collect the balloon basket and its passengers. Sometimes landings entailed those vehicles driving into or through farmers' vineyards and orchards in order to position the trailer correctly for the balloon basket to land on; although every effort was made to avoid damaging crops and trees, occasionally damage was inevitable. Being *elite balloons* and carrying significant prestige, they were generally forgiven and, if farmers were present to witness the landing, they would usually bestow hospitality and fresh fruit from their garden onto the balloons' pilots, crew, and passengers.

Circa 2009

2009: Elite balloons

Fieldnote Diary 1998: On a bus tour from the Turkish south coast to Cappadocia, as we enter the region, our tour guide promotes the opportunity to go ballooning at sunrise the following morning. I overhear the woman sitting behind me telling her friend that there must be some risks with flying in hot-air balloons. She explains to her companion that she *would* go ballooning if only she did not have young children. I later learn that 12 people on the bus have signed up to go ballooning.

At breakfast the next morning, the 12 balloonists arrive back and join the rest of the bus-group at their tables. They seem 'high' after their flight and rave about how spectacular it was. At the breakfast buffet, I meet the woman I had overheard as being too afraid to fly in a balloon. I ask how she is today. She replies that she is fine but that she "would be a hundred times better" if she had gone ballooning after all. Realising now that her fears were unfounded, her regret at missing out on an "unmissable" experience was strong. Meanwhile, the ballooning 12 are continuing to tell everyone in the breakfast room the details of their experience. The others listen on, envious, and seemingly regretful that they had not flown too.

Throughout the 2000s, Kaili and Lars continued to promote ballooning in Cappadocia, regularly attending international tourism trade fairs and inviting Turkish Tourism Board and international media personnel to take a balloon flight with them. Ballooning in Cappadocia increasingly featured in international newspaper and magazine articles as well as television travel documentary programmes. Michael Palin, of *Monty Python* fame, visited Göreme in 2007 for his *New Europe* television series and book, and while there, he flew with Kapadokya Balloons and had Kaili and Lars co-piloting; this took their international fame, as well as their elite status, up a further notch. By the late 2000s, Turkey's national tourism board was so heavily using images of balloons flying over the Cappadocia landscape in its international promotional campaigns that tourism in Cappadocia, and to some extent tourism in Turkey, had become synonymous with hot-air ballooning in Cappadocia. The cover photo used for the 2009 edition of the *Lonely Planet Turkey* guidebook was one of Kapadokya Balloons' two blue and yellow balloons flying over the fairy-chimney landscape. With hot-air ballooning becoming a "must-do" activity on tourists' itineraries, even those tourists who were otherwise holidaying on the coast were starting to make the long trip to Cappadocia for a day or two just to go ballooning. The increasing demand was met by a proliferation of ballooning operators; in 2008, there were seven balloon companies in the region, and by 2009, there were 11 companies operating. On most mornings during the 2009 summer season, there were between 30 and 40 hot-air balloons flying at sunrise over the Göreme area.

The strong demand led some companies to fit in two shorter flights each morning (ballooning must occur shortly after sunrise before winds and air currents become too strong). This meant that two 'classes' of balloon flight became available, with some companies offering cheaper short flights while others promoted their flights as a high-end, exclusive activity. The late 2000s thus saw a 'democratisation' of balloon tourism in Cappadocia, following a pattern similar to that which occurred for tourism in general in late nineteenth century Europe:

> The growth of such tourism represents a kind of 'democratisation' of travel. We have seen that travel had [previously] been enormously socially selective. It had been available for a relatively limited elite and was a marker of social status. But in the second half of the nineteenth century there was an extensive development in Europe of mass travel... Status distinctions then came to be drawn between different classes of traveller, but less between those who could and those who could not travel.
>
> (Urry and Larsen, 2011, p. 31)

The cheaper ballooning options were not only shorter in duration but they often involved larger balloons and baskets capable of holding as many as 30–35 passengers, limiting the ability of the passengers, crowded into the basket, to enjoy looking at the view over the basket's rim. These cheaper flights also tended not to include the fuller experience of watching and participating in the balloon's inflation and deflation, nor the post-flight champagne toast. Nonetheless, the cheaper flights still likely satisfied a touristic desire to do the "must do". Indeed, although not conducive to the sense of having an exclusive experience, the presence of large numbers of fellow passengers – both in your own balloon basket and in others – likely afforded what Urry (1990) termed the 'collective gaze'. According to Urry and Larsen (2011), the prestige afforded through this form of tourist gaze comes from the large numbers of other tourists who collectively "indicate that this is *the* place to be" (p. 19). Moreover, whether on a short or long flight, cheap or expensive, the passengers in a balloon basket became a part of the now-iconic spectacle themselves. With 30–40 balloons flying at the same time each morning, they were becoming quite a spectacle for tourists from the ground. Hence, even the larger balloons doing shorter flights afforded prestige to their passengers who not only became a part of the photographed icon, but also, literally and figuratively, looked down on the tourists on the ground who, again literally and figuratively, looked up at them.

Interestingly, although the 30–40 balloons flying each morning were considered a spectacle, tourism promotional material depicting ballooning tended still to show only one or two balloons flying over the Cappadocia landscape. This may have been due to it being still Kapadokya Balloons who were the main provider of marketing images; hence their one or two balloons tended to feature, such as in the 2009 *Lonely Planet Turkey* cover photograph. On the other hand, the depictions of more solitary balloons might also suggest that ballooning was still regarded as best promoted as an exclusive activity. This would encompass Urry's (1990) idea of the 'romantic gaze' through which, in contrast to the 'collective gaze', "tourists expect to look at the object privately or at least only with 'significant others'" (Urry and Larsen, 2011, p. 19); hence, "notions of the romantic gaze are endlessly used in marketing and advertising tourist sites, especially within the 'west'" (ibid., p. 19). That tourism promotions tended to show only one or two balloons flying over the Cappadocia landscape would suggest that, continuing up until the late 2000s, the *Elite balloons* should not be in crowds.

Paradoxically, however, the presence of several other balloons, plus the two 'class' tiers emerging at this time, enabled the elite status of the higher-end flights to be asserted in *their relation to* other balloons observed. As with the emergence of two classes of traveller in the late nineteenth century, status distinctions in Cappadocia ballooning became 'less between those who could and those who could not' fly, and more 'between different classes' (Urry and Larsen, 2011, p. 31) of balloon flight. Well into the 2000s, the "Classic flight" offered by Kapadokya Balloons maintained its position as the most expensive, and exclusive, flight available. As well as being a longer flight duration, and in a basket holding 8–10 people maximum, this *elite balloon* experience continued to involve watching the inflation of the balloon at the start and a champagne party upon landing. Following Kapadokya Balloons' *elite balloon* example, some newer companies – such as the aptly named 'Royal Balloon' company – similarly distinguished their offerings by promoting the promise of a longer flight duration and a limited passenger load in the basket. Some companies also started offering "VIP flights", which were private flights for two passengers only, priced at 600 euros for two.

In my 2008 fieldnote diary, I wrote about being in an *elite balloon* flight when the pilot pointed out the high number of passengers in another company's balloon flying nearby. He told us that the other balloon's basket takes up to 36 passengers; "a whole tour bus-load!" He then pointed out the two large tour buses parked at the other company's take-off site, exclaiming, "Imagine going on a tour bus to your take-off site!" The pilot went on then to lament what he saw as a drop in standard of balloon passengers: "Ballooning used to be a daring, sporty activity. Now we get these horrible people in stilettos arriving at the office going 'Where's the balloon'?" In this instance, the stilettos, as with the tour bus, became symbols of 'mass tourism'; in a similar way in which seaside resorts in England once became 'places of contested social tone and fights over cultural capital' (Urry and Larsen, 2011, p. 48), so had the balloons become sites of contestation over exclusivity and elite status. A short time later, it was observed by some of our *elite balloon* passengers that the two large balloons of the other company had landed again only a few hundred metres from their take-off site; their thirty-minute flight had more or less hovered over some scrubland and the road. In contrast, our *elite balloon* flight lasted for more than 2 hours and, travelling 12 kilometres, we had experienced a range of altitudes flying over several valleys and villages.

During the 2000s, again following the example of Kapadokya Balloons, many of the *elite balloon*s were flown by foreign pilots, hired by *elite balloon* companies because of a general perception of them as "better" pilots. Although they demanded higher wages than the now plentiful Turkish pilots, foreign pilots were understood to enhance their employer's reputation. Drawing increasing numbers of foreign pilots to Göreme, the *elite balloons* prompted the emergence of a new form of 'expat' circle. While a few of these (mostly male) pilots brought their family with them, most came as single men. They befriended other resident foreigners, seeking from them advice on how to live comfortably in rural central Turkey, as well as how to navigate relations with "local people". During this period, the pilots and other foreigners developed a particular 'expat' lifestyle; going to Göreme bars together, holding dinner parties, and sharing imported foods and recipes from home. The pilots tended to have wealthy backgrounds and an associated air of superiority about them. Their conversation frequently involved talking disparagingly about Turks, and particularly Turkish balloon pilots, who were also growing in number at that time.

Some Göremeli young men trained by Kaili and Lars went on to become successful balloon pilots and their occupation afforded them significant status among their peers as well as in the wider community. Previously, as was discussed in *Living with Tourism (1)*, the young men of Göreme had craved a life in a foreign country as a path to status and upward mobility. One of Kaili and Lars' trainees had two brothers who had gone to live in Australia, and the trainee might well have followed suit had he not had the opportunity of a prestigious career as a pilot within his hometown. Increasingly, other young men endeavoured to do pilot training too, with some raising the money to go abroad to obtain their licence. When they returned and began working for a balloon company, able to walk around in 'pilot uniform' of lapelled shirt and designer sunglasses, going to bars in the evening to woo tourist girls with their "I'm a pilot" talk, they developed a new confidence and status bestowed upon them by the *elite balloons*.

2009: Money-maker balloons

With the growth of the hot-air ballooning industry and the proliferation of balloon companies, 'commission' practice became well-established and wide-spread in Göreme during the late 2000s. The intense competition as more and more ballooning companies opened in the region led to their increased use of commission payments to compete for customers. In a 2005 interview I conducted with Kaili, she said:

> They are focused on making money out of ballooning now because there are more balloons and there is more capacity and there is more money in it… Everyone's after balloons these days, and therefore they are following the balloon trend and they're very alert to seeing anybody walking in the street with a [ballooning] brochure in their hand.

By "they", Kaili was talking about hotel and tour agency owners, tour guides, restaurant workers, and street "touts"; indeed, "anybody who is living, working locally". Referring to this as the "Göreme fishing net", she estimated that only 10% of Kapadokya Balloons' customers "managed to escape the fishing net which is laid out locally to catch people before they get here". The other 90% of customers made their booking "through somebody", and during the late 2000s, Kapadokya Balloons reportedly paid out over US$100,000 annually in commission payments.

With increased competition between balloon companies, the commission percentage of the flight sale price increased from 15% in the late 1990s to around 30% in the late 2000s. A few companies reportedly paid up to 50% or even 60% in commission, and some began to stipulate a fixed minimum price they should receive themselves so that touts could bargain with the tourists and vary the amount of commission they made on each 'sale'. A tout or a waiter may decide to add only €10 to the minimum price per ticket, but selling a balloon ride to a couple even at this rate was considered a decent addition to their otherwise low wage. Moreover, as more and more new ballooning companies were formed, the minimum price of flights was continually lowered to encourage more touts to find them customers. Hence, the money-making potential from ballooning increased and widened among non-ballooning businesses and individuals while, because of the substantial commission payments given out plus the reduced flight prices, the profits of the ballooning companies themselves decreased during this time.

The significant commission-earning possibilities from the *money-maker balloons* led to an increased fervour in the sense of business owners, and tour guides, having 'ownership' of their tourist customers, or guests. A hierarchy formed whereby, if tourists were travelling in Turkey on a guided tour, their guide would receive all the commission from ballooning and other sales. For tourists travelling independently of a tour group – which continued to be the predominant travel mode of guests staying in Göreme's accommodations – the hotel or pansiyon where they were staying would take primacy in the commission stakes. After that, the descending hierarchy would be approximately in the order of tour agency owners, touts attached to tour agencies, and then freelance touts, waiters, and others whom tourists may come in contact with. As Kaili described:

> It can be difficult because if they [tourists] do get scooped up by the tea boy or restaurant worker, they might have already booked by email, or they might already be booked through an agency or their pansiyon, and they are just on the road and trying to find their way here.

Ferocious arguments sometimes occurred when owners of pansiyons accused tour agency or restaurant staff of "stealing" their guests by selling them a balloon ride. Kaili continued:

> It is difficult…we then have to explain to the boy in the teahouse why he is not going to get paid sixty-five euros for that reservation, because the tourists had already booked from their hotel… and we have to prioritise the hotel as the source and the tea boy is pretty low down on the source ladder.

By the late 2000s, the average commission earning per ticket for hotel or tour agency owners selling balloon flights was €50; approximately 30% of a €170 ticket. When commission was claimed and paid, in an attempt to disguise what was being exchanged, it was generally referred to as an "envelope" rather than "commission". This attempt to disguise the practice highlighted the ambivalence felt towards the commission practices, despite their becoming so widespread; within an Islam-influenced moral economy commission was shunned as "*haram para*" (dirty money) by the Göreme community. Moreover, many Göremeli considered the high commission percentage paid on balloons tickets to be excessive and "not normal", saying that whilst a 3–5% "booking fee" was common in tourism around the world, 20–30% was "stealing". Nonetheless, it was the commission earned by freelance touts which was most strongly condemned, whereas that earned by the owners and workers of guest houses, restaurants, and tour agencies was seen as more legitimate, even if excessively high; after all, talking with 'guests' and making recommendations had always been considered a part of the business of tourism.

Additionally, from the guests' perspective, a flight purchased via their hotel was considered more trustworthy, and hence safer, than a cheaper flight being offered by a waiter or tout encountered in the street. If a hotel or pansiyon sold "quality" flights with a reputable balloon company, then word-of-mouth functioning among guests could guarantee further sales. As Kaili told me:

> I see the progression in our reservations and [name of hotel] always works with us because we have the best reputation…and especially from our 'Classic flights', we send them back so radiantly effusive and positive

about their experience that – guaranteed – if I have two passengers from [name of hotel], I am going to get two more the next day and two more the next day and so on through word-of-mouth.

It thus came to be understood that owning a pansiyon, or even better a 'boutique hotel', was the most effective way to make money from balloon-commissions. Hence, the number of accommodation businesses grew rapidly at this time, with people converting their house into a hotel or owners of existing hotels and pansiyons purchasing neighbouring houses in order to add more guestrooms. In addition, with the more upmarket accommodation more likely to attract wealthier tourists who in turn would more likely buy a balloon flight during their stay, many pansiyons were renovated to become 'boutique hotels' at this time, for example by converting what were previously dormitory rooms into more upmarket rooms with private bathrooms. A boutique hotel selling a double room at 100 euros per night could double its income from that room if the guests went ballooning. This 'boutique hotel' proliferation occurred predominantly in the older neighbourhoods, which consequently became dotted with construction sites as houses and existing hotels underwent conversions and renovations (see Chapter 5 for a more in-depth discussion on these neighbourhood changes).

Boutique hotels were not the only accommodation type embroiled with the *money-maker balloons*, however, with some lower-budget backpacker pansiyons being drawn into *money-maker balloons* practices also. This involved offering free accommodation for one or two nights, and then bumping the room or bed price up in order to encourage the guests to leave; the aim was to get as many customers through as possible, each for a short stay of only a night or two during which they would receive the 'hard-sell' on ballooning. With this approach, the money earned by the pansiyon was almost entirely from balloon-flight sales rather than from the accommodation per se. The owners of these pansiyons were accused by other Göremeli business owners of "not looking after customers and only looking at selling them balloons". Inevitably affecting the tourists' experiences of hospitality in their accommodation establishment, this practice was believed to reflect badly upon Göreme's hospitality (misafirperverlik) more broadly; a significant consequence, we could say, enacted by the *money-maker balloons*.

Overall, the spread of money-making possibilities and the broadening of the commission "fishing net" meant that a substantial portion of Göreme households came to earn at least part of their livelihoods from the hot-air ballooning sector during the period circa 2009. Furthermore, it was estimated that over five hundred people from Göreme and nearby towns were employed directly – as pilots, ground crew, drivers, and office staff – by balloon companies. Hence, because they were so instrumental in widespread money-making for a broad set of the population, from hotel owners, through direct balloon-operation employment, to otherwise low-wage earners in restaurants and teahouses, the balloons were tolerated even though a concurrent view of them as becoming a nuisance was developing. While I heard many complaints and negative views about the balloons, when I asked what Göremeli people's general thoughts about the balloons were, the reply was usually along the lines of: "They don't say anything because everybody's earning money from them".

2009: Nuisance balloons

Fieldnote diary 2009: Every morning I am woken early, sometimes by the pre-sunrise mosque call, but more usually these days by mini-buses calling at nearby hotels to pick up passengers going ballooning, or chase-vehicles darting at high speed through the narrow streets of the old neighbourhood with the clattering of balloon trailers they are towing bouncing on the cobbles. Some mornings I wake to a strange hum which at first I could not figure out – not loud but a constant hum hanging in the air sounding like a gigantic swarm of bees is approaching Göreme. I then realise the hum is caused by the great many industrial-sized fans blowing air into the balloons to inflate them at their take-off sites out in the valleys... I fall back to sleep for an hour, until I wake again, this time to a whoosh sound, outside but close by. I get up and open the curtains to look out. There, right outside my window are 20 tourists in a balloon basket. They see me and wave. Some point cameras at me. I snatch the curtains to close them. I had not expected these eyes and cameras prying into my second-floor room.

What I will call the *"nuisance balloons"* emerged during the late 2000s as a new balloon configuration, or 'character'. The *nuisance balloons* drew together a broad array of people, places, and things, including vehicles and trailers, fans and gas, gardens and crops, as well as farmers, women, and other villagers who otherwise were not connected directly with the balloons or tourism business. Along with the proliferation of balloon companies and the rapid increase in the number of balloons flying each morning, a sense of resentment towards the balloons appeared to build throughout the 2000s: during my fieldwork in the last years of the decade, the balloons were the topic of a large proportion of conversations I had with Göreme people. Everyone commented on how many balloons there were, and many said they had come to "hate the balloons". I was told by one woman: "Nobody likes the balloons! They look beautiful, but they are bad!" The increasingly negative mood in relation to the balloons was at least in part linked to the *money-maker balloons* which, despite the employment opportunities the balloons provided, had become associated with jealousy and fighting, as well as a sense of moral corruption associated with *"haram para"* (dirty money) commission payments.

In 2008, the Göreme Mayor imposed restrictions to prevent the balloons from being flown low over Göreme. This move was ostensibly made in order to reduce the noise disturbance created by the balloons very early each morning. In addition, the ever-growing number of balloons flying over houses and hotels meant that they had become a nuisance due to their becoming an invasion of privacy. A hotel owner told me that everything was okay a few years ago when there were just 1 or 2 balloons flying in the area, but with over 30 balloons flying each morning, it was not uncommon for himself or his guests to wake up in the morning with balloon passengers looking through their upper storey window. Balloon passengers were also able to look down into, and photograph, the otherwise enclosed courtyards attached to villagers' houses. These courtyards were normally secluded behind a high wall in order to protect the female virtue and honour of the household (see Delaney, 1991); by flying above villagers' houses, the balloons circumvented this 'enclosure', thus potentially compromising the virtue and honour of the household.

The increasing numbers of balloons flying each morning had also begun to raise concerns regarding the possibility that the balloons were emitting dangerous gases, and that these gases might have detrimental effects both on villagers' crops and the fertility levels of young women. Stories circulated that this was a further reason for the low-flight ban; it was to prevent the balloons with their "poisonous gases" from flying too close to villagers' gardens and women. There was fear at that time of Göreme losing its 'municipality' status due to the drop in population associated with displacement of households from the old neighbourhoods as houses were converted into hotels (see Chapter 5). If Göreme's population were to drop down to 'village' level, it would lose its municipal governance council. The links made between the balloons' "poisonous gases", fertility, and this population issue, while tentative, were indicative of increased negativity felt towards the balloons. The balloons' gases were also blamed for the dying trees and vines in villagers' orchards and vineyards. For example, the woman quoted above who said that the balloons were "bad" went onto tell me she had heard that the balloons' gases were killing the grapevines and turning their leaves black: "I don't know if it is true but I have heard that", she said. With the correlations between women's fertility and that of fields, or soil, drawn in Turkish village society procreation beliefs (Delaney, 1991), it is little wonder that if the balloons' gases were believed to affect one, then they would also be seen as affecting the other.

Whilst theories about the damage caused by the balloons' gases remained unconfirmed, there was little doubt that the balloons caused damage to gardens and crops more directly, especially when they came down to land. For example, at times, the top branches of apricot or apple trees in orchards were skimmed over in order to lower the balloon's speed before landing. Moreover, with 30–40 balloons flying each morning, the associated 4×4 vehicles and trailers driving onto and across fields in order that the ground crew and trailer were in position when the balloon landed could cause considerable damage to the fields and crops. Kaili and Lars told me that, while there was an unwritten rule among the ballooning fraternity that they should be careful of crops, increasing numbers of pilots in Cappadocia were not careful in this regard. Although it was still common for farmers to show considerable hospitality to landing balloons and their passengers, the number of incidents where anger was shown towards the balloons upon landing were increasing at this time; I heard reports that on several occasions, villagers and farmers shouted and waved their fists at balloons.

The increasingly contentious use of orchards and vineyards as take-off and landing sites was reduced in part by some balloon companies buying gardens or orchards from villagers to use as permanent take-off sites. However, the landing site could not so easily be controlled, and so the potential for damage to be caused by the balloon vehicles and the passenger and crew foot traffic upon landing remained. Consequently, as well as implementing the low-flight ban over Göreme, the Göreme Mayoral/Municipality office (*Belediye*) set up a "*koruma*" (security or protection) watch, employing local men to follow the balloons and take photographs of any misdemeanours being committed, particularly in relation to gardens and crops. Nuisance behaviours they looked out for included balloons causing damage to trees, balloon vehicles driving through gardens, and pilots allowing tourists to help themselves to grapes or apricots after landing. They also reported any balloons going below the low-flying ban level when flying over Göreme's neighbourhoods. If caught committing any of these violations, the balloon company involved would receive a fine, although the companies' awareness of being observed was fairly effective in itself in regulating at least some aspects of the nuisance balloons.

2009: (Un)regulated balloons

During a conversation, as recorded in my 2009 fieldnotes, I had with Mehmet (the man quoted above as likening the balloons to the big fish in *Jonah*), he told me he imagined that "soon the sky will get closed with balloons so that when you're in a balloon you won't see Cappadocia any more, you'll only see balloons". "They need to stop them", he continued, "because not only is their gas killing our trees, but there are too many cowboy companies opening and they'll have accidents. They need to restrict their licensing more; they need to be more strict!"

The idea that "the sky will get closed with balloons" was already, in part at least, becoming a reality: A hotel owner told me that on some mornings, the more picturesque valleys became "totally crowded" with balloons jostling for space. Like Mehmet, the hotelier lamented the fact that there were no aviation regulations to limit the number of balloons flying in a particular space, and he, too, predicted that more accidents would happen. While some accidents had already happened, they did not appear to have prompted civil aviation authorities to increase regulation. It was therefore the *un*regulated nature of the balloons which appeared to cause the most concern at this time. This included a reportedly loose pilot accreditation system whereby young men from Cappadocia were going to Russia to obtain, relatively cheaply and easily, a balloon pilot licence. Regulation concerning the suitability of weather conditions for flying also remained largely absent. In 1999, as set up by Kaili and Lars at that time, determinations about the wind conditions had been at the discretion of the pilots. A decade later, however, this judgement was more often made by the balloon company owners; I was told by some pilots that they were pressured by their company to fly even when they felt the conditions were unsuitable.

Circa 2019

2019: Elite balloons

'Kapadokya Balloons is the first private, licenced and commercial passenger hot-air ballooning company in Turkey. The company's pilots' superior experience in the region allows them to choose the best launching locations for the most scenic flights possible for each day's weather conditions. The fleet of luxury aircraft are the newest, safest and most comfortable balloons available in the world today. Kapadokya Balloons' insurance safety standards far exceed that required by the Turkish, European and American standards. Having a professional managerial staff and experienced ground team, the company guarantees 100% customer satisfaction. Over 21 travel guidebooks and independent travel writers recommend Kapadokya Balloons for high quality flight experience and general services' (http://doraktour.com/assets/pdf/gold-catalogue.pdf)

When the Kapadokya Balloons business was sold and Kaili and Lars left Cappadocia at the start of the 2010s, in large part, the *elite balloons* left with them. This was despite attempts by the new owners to ride on the coat-tails of Kaili and Lars' legacy, as illustrated in the above extract quoted from the website of the Istanbul-based tourism company which had bought Kapadokya Balloons. There is no doubt that other remnants of the *elite balloons* lingered on also. People who had worked for Kapadokya Balloons when it was owned by Kaili and Lars fondly remembered the 'heyday' of the *elite balloons*. Other friends and associates I spoke to in 2018–19 also reminisced about those old days, such as elderly 'Bicycle Mustafa', so nick-named because he used to come by the Kapadokya Balloons office on his bicycle and sell his organic tomatoes. A trace of Kaili and Lars' *elite balloons* lingered on, also, on a restaurant menu. The restaurant directly opposite the Kapadokya Balloons 'hub' had retained a dish named "Lars Special" on its menu so that, every time a customer would ask what the dish was, the restaurant's owner was given the opportunity to tell of his long acquaintance and friendship with the first balloonists in Göreme.

With 28 ballooning companies and 240 balloons registered to fly in Cappadocia, however – plus many of those having ever larger 'mass tourism' baskets –

the elite status that had surrounded the balloons two decades earlier had largely dissipated by 2019. That ballooning had grown to an industrial scale now required an industrial-sized workforce, and it was estimated that approximately 200 balloon pilots worked in Cappadocia in 2019 (Özgentürk, 2019). While an increasing proportion of these were graduates of the three pilot-training establishments now set up in the region, many pilots came from other parts of Turkey to seek employment in Cappadocia's ballooning industry. I was told by one Turkish pilot I chatted with in 2019 that few foreign balloon pilots were employed in Cappadocia, mainly because foreigners tended to demand higher pay (reportedly upward of 4,500 euros per month), as well as necessitating the organising of a work visa. The monthly salary for Turkish pilots was generally a lot less (between 1,600 and 1,800 euros), plus, the pilot told me, the balloon companies could get away with less favourable work conditions in general for Turkish pilots: "That is why the balloon companies generally prefer to employ Turks these days", he said. Having moved to Cappadocia a few years earlier from the west of Turkey where he had piloted gliders and microlights, this pilot appeared rather jaded, saying: "It looks like a good job, but in reality it is not so good". For example, even though the Turkish Civil Aviation Authority had implemented the rule two years earlier that pilots could work a maximum of six days in a row and must have the seventh day off, they tended to have no regular, planned day off. With pilot-training institutes producing new graduates all of the time, moreover, there was concern among pilots that an over-supply would mean that wages and conditions would continue to go down. The glamour appeared to have been largely stripped from what used to be considered an elite occupation.

Figure 3.2 Tourists and balloons, 2019.

For the reportedly half a million passengers now flying yearly in the region (Özgentürk, 2019), however, while the air of personal exclusivity surrounding ballooning had largely gone, flying in a balloon in Cappadocia continued – more than ever – to be a sought-after, "must-do" activity. In 2019, Tripadvisor listed ballooning in Göreme in its top ten tourist experiences in the world and it tends to appear quickly in a Google search of the world's most popular 'bucket-list' items. Throughout 2019, with usually more people wanting to fly than spaces available each morning, many were prepared to pay high prices for even the shorter flights. The increased attention on social media platforms no doubt had much to do with this. Indeed, in 2018–19, the status attainable through association with the balloons had become so heavily incorporated into the social media 'like economy'[2] that the *elite balloons* of old had now mutated into a new balloon character: the *Instagram balloons*.

2019: Instagram balloons

> Technological tools are not passive objects; they carry affordances enabling as well as shaping tourist behaviour. A tourist with a smartphone in her hand is a hybrid with a number of new action possibilities.
>
> (Munar and Gyimóthy, 2013, p. 246)

Notes from 2019 fieldwork diary: At 5.30 a.m. this morning I went up to the roof terrace of a neighbouring hotel to observe the *Instagram balloons* in action. I counted around 60–70 balloons visible in the sky above Göreme, some closer and some further way in the direction of Rose Valley. Against the light of the rising sun, I could make out the silhouettes of hundreds of people moving around on 'Sunset Point', over which many *Instagram balloons* were flying so low as to almost touch the ridge.

The roof terrace I was on was decked out with Turkish rugs and cushions. A low 'sofra' table was laid with a photogenic 'breakfast' of colourful fruit, teapot, wine glasses, and cherry juice. Various groups of tourists took turns in sitting at the 'breakfast' table for photos with the *Instagram balloons* behind them. The two hotel personnel on the terrace managed the queue for the mock-breakfast table photograph spot, and they also helped the groups by taking photos of them.

Hotel guests on the terrace included:

A group of four (two couples) from Brazil: Taking selfies as well as photos of each other in various combinations of the four and in various poses, and doing video-blogs twirling around to music – always ensuring the balloons were in the background.

Two Russian women: In long floaty dresses and with long floaty hair photographing each other.

A Thai couple (currently living in Manchester, England): Set up their camera on a tripod to do time-lapse photography of the *Instagram balloons* floating past.

Two Chinese couples: Although apparently not together, both women were similarly dressed in 'boater' hats and long dresses – one red and one white. The men photographing their partner in various poses (mostly with their back to the camera and 'looking out' at the *Instagram balloons*).

An English/Australian couple: Sit on the cushions playing with the hotel's puppy. Periodically one of the pair stands up and takes a photo of the balloons.

A Chinese couple: Come up the steps onto the terrace. The hotel staff don't recognise them and approach them to check if they are staying in the hotel. They explain they are staying in a neighbouring (partnering) hotel and had been told that they could come up onto this terrace. They are allowed onto the terrace to begin their photoshoot.

Looking out over the old neighbourhood behind me, I can see similar behaviours repeated on numerous hotel roof terraces. A few of the terraces have heart-shaped metal frames, some with a swing hanging in the frame, for couples to add an extra touch of romance to their *Instagram balloon* posts.

I was just about to leave when a couple came up onto the terrace. They were from Kazakistan. They told me they had got up at 4 a.m. to hike to Rose Valley to photograph the balloons there. They showed me the great many photos they had taken there – many were of her posing in front of the balloons. He had been to Turkey two years ago, to Istanbul and the coast. He said he had not even heard about Cappadocia then; "Now, though, Göreme has become the No.1 spot in Turkey – All because of *Instagram*", he said. He asked me if I knew anywhere else in the world with a comparable amount of ballooning. I told him I had heard that Cappadocia was the biggest ballooning centre in the world, and that I had read in a Turkish newspaper that over half of the entire world's hot-air ballooning now occurred in this region. He said, "It's incredible to have such a balloon festival EVERYDAY here!" He continued, "It is all about *Instagram* here now. There are so many images of the balloons on there".

Impressed by the photographs the Kazakistan couple had shown me, the following morning I decide to follow their lead and head out before sunrise towards Rose Valley. Beyond the edge of Göreme I come to the area – near Abbas' vineyard (the one where, he had told me, the vines' leaves cannot breathe) – where there are multiple balloon take-off sites.

It is quite a magnificent scene of maybe 30 or 50 balloons being lit up in the darkness by their gas flames. Crowds of tourists in baskets whoop excitedly as their balloon leaves the ground and floats upwards and away. As dawn breaks and the area around me becomes visible in the morning light, I realise that there are as many excited people on the ground as there are floating overhead in balloon baskets. There are several mini-buses and an indefinite number of taxis parked all over the area, having brought the hundreds of *Instagram balloon* watchers to the area. People are running all over the hilly ground, trying to get into optimum posi-tions when balloons are flying overhead. There are props of vintage cars being sat in or stood on, women in long flowing dresses, and couples in wedding outfits, and there are photographers with tripods and lamps. The whole valley and the sky above it has become a giant photo-studio, in which the *Instagram balloons* are the most important element.

The 'tourist with a smartphone in her hand' (Munar and Gyimóthy, 2013, p. 246) plus the introduction of social media as a major global actor led, during the 2010s, to the emergence of what I will characterise as the *Ins-tagram balloons*. Whilst tourism had become increasingly mediated through social media more generally during this decade, after its launch in 2010, the image-sharing mobile-phone application *Instagram* became one of the fastest-growing social media platforms globally (Smith, 2018). As 'a uniquely fecund space' (Smith, 2018, p. 175) for images of places to be circulated meant that the 'surging popularity of the platform, and its deep imbrication with the mythos of travel, mirrors the precipitous expansion of the tourism industry itself' (Smith, 2018, p. 173). In this sense, *Instagram* can be viewed as a significant factor propelling both the rapid expansion of Göreme's hot-air bal-looning fame and the rapid growth of tourism generally in Göreme during the 2010s. As the Kazakistan man above said, at the time of his visit to Turkey in 2017, he had not even heard of Cappadocia, nor its balloons. Yet remarkably, by 2019, Göreme had become the "No. 1 spot in Turkey", and this was "All because of *Instagram*".

I explained in Chapter 2 how local hotel owner Mehmet undertook various measures – particularly using social media as an 'influencer' – to attract new markets to his hotels. Having part ownership in a balloon company, also, he endeavoured to create a circulation of images which would further promote not only his hotel but the idea that Göreme was *the* place for hot-air balloons and ballooning. Mehmet initiated the idea of decorating a roof terrace at his hotel with Turkish rugs and a 'breakfast'-adorned table, so that his guests could photograph themselves as if having breakfast with the sky filled with balloons in the background. It did not take long before pictures taken on the hotel roof terrace became as frequently posted on the popular Chinese travel-oriented social media platforms as images of tourists actually flying in a balloon. After the mid-2016 attempted coup and the tourism slow-down which followed, Mehmet embarked on a drive to 'influence' the Russian market, particularly, by inviting groups of Russian models. He also invited many globally recognised social media "influencers" to stay as his guests and to take a balloon ride with his company. In addition, Mehmet learned how to 'play' *Instagram* for optimum effect; using "hashtags" so that images of his hotels would be viewed, not only by his 200,000 followers, but also by multiple millions of followers of other *Instagram* accounts. As the man from Kazakistan said: "It is all about *Instagram* here now."

As the *Instagram balloons* emerged as a new balloon configuration during the 2010s, they gave rise to and incorporated a whole new set of practices, places, people, and businesses. Firstly, the way in which hotels were incorporated in with the *Instagram balloons* practices network departed substantially from the ways in which Göreme's hotels were previously tied in with the balloons as *money-makers*. In short, with the emergence of the *Instagram balloons*, a hotel's location and ability to provide a roof terrace for good balloon-viewing became primary. At Mehmet's hotels, every morning throughout 2018–19, there would be a queue of guests on the roof terrace waiting to have their few moments on the cushions next to the decorative 'breakfast' table. Many hotels in Göreme followed Mehmet's lead and created a decorative roof terrace – some of which included metal heart-shaped frames – specifically for balloon-viewing and *Instagram* photography. While hotels in the upper parts of Göreme's neighbourhoods already had an advantaged, elevated position, some hotels in lower areas added an extra floor in order to create a vantage point, or promontory, over neighbouring hotels. Consequently, a race developed whereby more and more floors and roof platforms were sequentially added to neighbouring hotels. The hotels also engaged in a race for 'likes' on *Instagram*, with some hotels in the upper parts of the village – particularly those close by Mehmet's hotel which had the original *Instagram* terrace – being booked up entirely throughout 2019. Hotel reservation staff told me that, before booking, guests always enquired to confirm that they would be able to "have breakfast with the balloons" on the hotel's terrace at sunrise.

While flying in a balloon undoubtedly afforded passengers a promontory and distanced position from which a spectacular view of the Cappadocia landscape could be gained, the balloon basket – being often crowded and cramped – was not necessarily the best place to get *Instagram*-worthy photographs. Indeed, only a small portion of images of Cappadocia balloons would appear to be taken in or from a balloon basket. Rather, the majority of *Instagram balloon* images are taken from on-the-ground promontories. Along with the hotel roof terraces, other promontories that became popular *Instagram balloon* spots included the area towards 'Rose Valley' just north of Göreme, and 'Sunset Viewpoint', a ridge above Göreme's old Gaferli neighbourhood. Sunset Viewpoint had already become popular as a place for tourists to go in the early evening to watch the sun setting over Göreme, and the Göreme Municipality Office (*Belediye*) had set up a rope barrier around the cliff-edge as well as erecting a wooden platform over the most popular promontory point. Additionally, the owner of the land just next to this promontory built a café and shop, as well

Figure 3.3 Traffic jam to access Sunset Viewpoint at *Instagram balloon* time.

Figure 3.4 Tourists on Sunset Viewpoint ridge at sunrise.

as putting picnic seats around the area for people to sit and have drinks and ice-cream while they watched the sunset. During 2018–19, the early mornings also became a highly popular time to visit this ridge area, since increasing numbers of well-'liked' images circulating on *Instagram* identified it – particularly for those not staying in a hotel with a good viewing terrace – as a good spot for interacting with the *Instagram balloons*. In the summer of 2019, the narrow road leading up to the ridge became so chaotic at both sunset *and* sunrise time, that the Göreme Belediye decided to close the road to cars – as well as to charge people a small ticket fee to walk up to the ridge – at these times of day.

Each morning, the whole Sunset Viewpoint ridge area would become crowded as hundreds of people engaged with the *Instagram balloons*, many of which would fly low over the ridge to give an extra thrill both to passengers and photographers on the ground. While one side of the ridge looks down over Göreme's neighbourhoods, in the other direction, a fairy chimney filled valley (known locally as "Love Valley") afforded tourists good opportunity to take post-worthy 'selfies' or 'promontory witness' images with the *Instagram balloons* and Love Valley in the background. Sean Smith (2018, 2021) discusses the 'promontory witness' pose as a common *Instagram* visual motif; usually taken from a high vista and involving the tourist – whose back is usually turned to the camera – as witness figure looking upon a sublime landscape. Smith argues that, with the tourist depicted as the privileged *sole* consumer of the sublime landscape – which in this case also includes the balloon-filled sky, such imagery is deeply imbricated with discourses of power and self-aggrandisement in its 'alluding to mastery of the scene beheld' (Smith, 2018, p. 180).

There are significant parallels between the *Instagram balloons* and some earlier aspects of the *elite balloons*. Despite there being crowds of other tourists just out of view on Sunset Viewpoint ridge, 'promontory witness' images would tend to depict the place as empty except for the 'witness' themselves. The elite balloons, in their early days at least, tended similarly to be depicted as having *lone* mastery over the landscape. Conversely, however, the *Instagram balloons* are, by necessity, *many*. Indeed, it is the sky filled with balloons that fits so well with the Romantic sublime aesthetic Smith (2018) argues to be prominent in *Instagram* imagery. Hence, it was the great many *Instagram balloons* flying over Göreme each morning which not only led to Göreme becoming thoroughly implicated in the 'like economy' (Gerlitz and Helmond, 2013), but also to Göreme's becoming the main tourist centre of Cappadocia.

The Instagram balloons also significantly impacted, or *acted*, both on how tourists interacted with different aspects of Göreme as a destination, and on how businesses in Göreme came to service those tourists' interests. Many tourists' choice of hotel started to be based on a hotel's 'promontory' position and the view of the balloons afforded from its roof terrace. Taxi's and mini-bus tours took tourists around the well-known *Instagram balloons* spots in and around Göreme, so that in the early mornings, these places became over-run with tourists and their cameras, or their hired photographers, along with tripods, selfie-sticks, and a milliard of visual props. Many new types of businesses opened, such as those hiring out vintage cars, wedding dresses and other costumes, carpets, and professional photographers. Other types of businesses, such as bars and discos that used to operate into the night closed down, declining in popularity due to the early morning rise necessitated by the *Instagram balloons*.

2019: Money-maker balloons

The way in which the balloons acted as money-makers changed considerably during the 2010s. This had a lot to do with the Istanbul-based company referred to above that had purchased Kapadokya Balloons almost a decade earlier. By 2019, the company had ownership of three further ballooning operations and, as well as these, since 2017, it had taken out lease agreements on 15 of Cappadocia's other ballooning operations. The company thus became not only the single biggest player in the ballooning industry, but it gained significant control over the ballooning economy overall. In June 2019, this monopoly position was reported in the *Hürriyet Daily News* as follows:

> A company's consolidation of some 19 hot air balloon companies has come under scrutiny for breaking down free and fair competition for the popular rides over the Cappadocia region's spectacular rock formations. By establishing such a consortium… the company thereby enjoys a monopoly, say many regional tourism actors… The Turkish Competition Authority has got involved and is currently looking into the situation. Some tourism players have said that balloon companies that are not part of the consortium have been prevented from providing the service. There are 25 registered balloon companies in the region, 19 of which are members of the consortium.
>
> (Özgentürk, 2019)

It was reportedly during the period following the mid-2016 attempted military coup and tourist numbers dropped dramatically in Turkey that this company moved in to offer the region's ballooning operations the apparent lifeline of a guaranteed income for a three-year period. With so few tourists coming to Göreme, only a handful of balloon operators declined the offer. Throughout 2017, most of the operators who had joined the consortium likely felt that they had done the right thing. However, as tourist numbers rapidly picked up again in 2018 and even more so in 2019 – and especially with the advent of the *Instagram balloons* – it became a regrettable decision. The 15 ballooning companies which had 'leased' their balloon basket spaces found themselves tied to a contract which severely limited their profit-making capacity. Some back-of-an-envelope calculations conducted by Göremeli hoteliers and ballooning industry personnel I spoke to suggested that, from their approximately 1,800–2,000 balloon 'seats', the Istanbul company's daily takings from ballooning were somewhere around the quarter-of-a-million euros mark.

Furthermore, as many tourists on shorter stays were desperate to get a ticket, the Istanbul company reportedly created a virtual auction system whereby tickets would sell to the highest bidder. This could push the ticket price up very high, especially after a run of bad weather, with some tourists apparently paying three times the usual price (500–600 euros instead of 200 euros) for a balloon ride. As a result, often the different people in one basket had each paid vastly different prices for the same flight; if those passengers talked to each other and learned how much each other had paid, they could be left feeling very angry about being "ripped-off". Another related "rip-off" was the emergence of internationally operated fake websites which purported to sell balloon tickets online; when the tourists arrived in Göreme, they found the balloon company and the flight they had bought online tickets for did not exist. Göremeli people felt very concerned about these various "rip-offs" and worried that tourists would assume it was they who were ripping people off. One man told me: "The tourists think it is us ripping them off. It reflects on us, on Göremeli. But it is not us, it's not connected to us. It makes us embarrassed, and ashamed". In 2019, there was a sense that, with external parties moving in, the balloons as moneymakers were falling away from Göremeli people's hands.

Largely due to the Istanbul company's monopoly control, the commission-payment system which had been so rampant a decade earlier had more or less finished: "Commission from balloons is dead now", one hotelier told me. He went onto explain that the Istanbul company did not work to a commission system; instead, the hotels had to buy a bulk number of balloon tickets in advance at the start of the year. Some hotels paid 50,000 euros to guarantee a certain number of balloon tickets for their guests throughout the year. However, as the tourism popularity of Göreme in 2019 surpassed everyone's expectations, the number of tickets hotels had purchased tended to be too few and there always seemed to be a scramble to find tickets for guests, again frequently pushing the price of the tickets up. For some smaller hotels and pansiyons, the outlay of a bulk-buy of tickets at the start of the year was prohibitive, plus they were wary of the risk of an unexpected downturn in tourist numbers. Many hoteliers told me that, consequently, they no longer bothered to try to book balloon flights for their guests since it was now too difficult to find places for them.

The popularity and fame of the balloons and the increase in tourist numbers they brought to the region inevitably led to many other balloon-related money-making opportunities, both directly and indirectly. The balloon sector continued to be a major employer in Cappadocia, with many local people working as pilots, crew, drivers, and office staff. While the hotels had turned away from earning commission from the balloons, they instead focused their efforts on linking in with the *Instagram balloons*, for example, by decorating a roof terrace or building a balloon-viewing platform. Some hotels – such as the one converted from Abbas's family's old cave-house situated directly underneath the Sunset Viewpoint ridge – were able to promote their location's easy access to Sunset Viewpoint, thus using that to boost their bookings and increase their room prices. Restaurants and other tourism businesses also inevitably benefitted from the increased balloon-related fame of Göreme; balloon-themed souvenirs and other paraphernalia multiplied, including balloon-themed ceramics and textiles, lamps, and keyrings. Some entirely new products connected to the *Instagram balloons* emerged. These included professional photographer services, early morning '*Instagram Tours*' taking people around key promontory spots to photograph balloons, and businesses hiring out props for *Instagram* photography, such as Turkish rugs, 'traditional Turkish' costumes, bridal outfits and other long dresses, and vintage cars. Many of these businesses also engaged the commission system to boost their sales, for example, hotels or agencies would earn a high-percentage commission for booking their guests a two-hour vintage car rental at 120 euros. Hence, commission could still be earned indirectly from balloon-related businesses even if commission payments from balloon ticket sales no longer occurred.

2019: Nuisance balloons

A 2019 article in *Al-Monitor* cited statistics which showed that 30,514 hot-air balloon trips were conducted in the Cappadocia region in 2018. The article quoted a planning scholar as saying, in response to this figure:

> They carried 537,500 visitors in these balloons. This is on average 2,500–3,000 people per day. Now think about all the buses, cars used daily to shuttle these passengers. Think about how much propane gas is used for the balloon rides? For the balloon to take off properly they cut several trees, and had to flatten the surface. According to our independent measurements over 200 soccer fields' worth of area was flattened. Gradually, the hot air balloon trips have destroyed the natural flora and fauna of the area.
>
> (Tremblay, Al-Monitor, 2019)

In 2019, complaints about and cursing the balloons was plentiful among Göremeli people: "Bloody balloons!"; "I'm fed up with the balloons – every morning they wake me up – they should be banned!"; "There are too many balloons. I'm sick of them! And I'm sick of tourism! It's too much!" One Göremeli businessman told me that the balloons had "messed with relations in Göreme". He continued:

> These days you go to see your friend and instead of asking 'How are you?' or 'How's your family?', they just say 'Oh no, the balloons are cancelled tomorrow' or 'I need to find more balloon tickets for tomorrow'. The balloons are everything now – they're the whole conversation.

Whilst in 2009 the Göreme township had seemed, to some degree at least, to have taken matters regarding the *nuisance balloons* into their own hands through the imposed low-flying ban and the "*koruma*" (security) patrol observing the balloons and passengers in the mornings, over the next decade there was an increasing sense of a loss of control. In 2010, reportedly, the newly elected Mayor of Göreme had lifted the low-flight ban so that pilots were able once again to fly as low as they wished over houses and hotels. This move was likely encouraged by the emergence of the *Instagram balloons* and the desire to fly low over the various *Instagram* promontories, including hotel roof terraces. While safety concerns led to the low-flight ban being reinstated later in the decade, the *Instagram balloons* frequently broke the rule but mostly went unreported. One evening I sat chatting with Abbas, Senem, and their neighbour, and the neighbour complained about the balloon noise that morning: "Firstly the fans were very loud this morning and woke me up", she said, "and then they all flew low over our house, and it was so noisy". Abbas told her to complain to the *Belediye* (Municipality Office). He said, "the balloon companies are supposed to be fined 30,000 tl if there is a complaint about them flying too low over houses". Everyone then agreed that, since the fine never seemed to be imposed anyway, it was not worth complaining. A few days later, when in conversation with another man who told me about being "sick of the balloons", he took out his phone and said: "Look, I took this video through my bedroom window this morning when I got woken by the noise from the balloons". The short video showed many balloons flying right next to his house.

The new Mayoral office had also seemingly given up on the "*koruma*" patrol, and the impact on gardens and crops of the ever-increasing number of vehicles and people associated with the balloons had become significant. Many gardens and orchards had been bought by balloon companies to turn into take-off sites. Consequently, the many vehicles associated with transporting balloon passengers, plus the taxis, minibuses and vintage cars associated with people going to photograph the *Instagram balloons,* created so much dust that they severely damaged any remaining grapevines and other crops. Moreover, in the afternoons and evenings, these now empty spaces became playgrounds for hundreds of tourists on ATV, or quadbike, tours. One man who still owned gardens told me: "All those empty spaces create too much dust. All the fields are getting dry – they are not green anymore, because of the dust and the gas! They [the balloons] shouldn't be allowed in Cappadocia anymore". Abbas' gardens located adjacent to a number of balloon take-off sites were severely affected by dust created by the balloon-related vehicles in the mornings and the great many ATVs taken to the area when the take-off sites became "empty spaces" in the afternoons. Indeed, so many tourists were coming to the area that Abbas had erected a high fence with barbed wire around his most prized vineyard, he said to prevent people from eating his crops. The fence could not keep out the dust, of course, and so it was ultimately perhaps a futile defence against what was by now a seemingly unstoppable barrage of the balloons and the other aspects of tourism the balloons brought with them.

Figure 3.5 Previously vineyards and orchards, now balloon take-off and carpark sites.

2019: (Un)regulated balloons

Extract from fieldnote diary, 2019: In conversation with a Göremeli hotel and tour agency owner, he told me:

It is dangerous because the sky is too crowded. I have 4 relatives who are pilots and they say it is dangerous. One of them told me that there was one time when he couldn't go up, or down, or any direction, because there were balloons all around him, and he was going to crash into a fairy chimney, so the only thing he could do was to land. So he did, in the road, in *Aydınlı* neighbourhood. And he got fined for landing there, even though he told them "If I hadn't landed, we might have killed 24 people!" So, the pilots don't like flying here anymore because it is too scary – too dangerous. Some pilots don't behave properly – the one above can look down so is supposed to think about the balloon below and go up to give them space if need be. But they don't always do that.

During the early part of the 2010s, although the continually increasing number of balloons meant that the sky was becoming "too crowded", the balloons had remained largely unregulated. As the decade went on several serious accidents occurred. Indeed, accidents became so regular that the hospitals in nearby Nevşehir reportedly created contracts with ballooning companies to allow their ambulances to wait and be the first to arrive when an accident happened; "It was a crazy period, with many of the pilots ignoring the rules and bumping into each other, so accidents happened very regularly", a pilot told me. He went on to explain that accidents were usually due to "the crazy young pilots not paying the proper attention to the other balloons in the sky around them".

The frequency with which accidents were happening prompted the Turkish Civil Aviation Authority to impose a limit of 100 balloons taking off at sunrise, with 50 further balloons able to take off half an hour later when some of the first batch were starting to land. While this limit alleviated some problems, many other aspects of ballooning remained unchecked, such as monitoring of weather conditions for flying, licensing and experience levels of pilots, take-off and landing processes, and basket-passenger load numbers. Consequently, even after the Civil Aviation Authority set the 100 + 50 balloon limit, serious accidents continued to occur, with several ballooning fatalities and serious injuries occurring between 2013 and 2017 (Aslaner, 2019; Erbil, 2017). When I asked a pilot why this was the case, he said in relation to the 100+50 balloons limit: "It is still too many – in the limited space – because they are too close together, so it is still dangerous. You have to constantly look below, above, around for other balloons".

As the decade went on and accidents continued to happen, a group of 'head pilots' lobbied the Civil Aviation Authority to step up their regulation of ballooning in the area. As a result, in 2018, an air traffic control system was implemented to monitor meteorological conditions, so that, as one pilot put it, "Now they have to give the green light before we can fly". This pilot went on to explain that the flying conditions in Cappadocia were very specific, partly because of the difficult fairy chimney terrain, but also because the sky is so crowded with balloons; "The conditions are quite different here", he said,

> so pilots have to be trained to the specific conditions of Cappadocia. They have to learn to be nimble with the balloons, to be able to safely fly low over and around the fairy chimneys, and they have to constantly look out for the other balloons around them.

He said that in other parts of the world, with flat terrain and far fewer balloons in the sky, it is possible to fly safely in 50 km/h winds. In Cappadocia, the new weather-related regulations were set with considerable caution so that when winds were 24 km/h or above, ballooning was cancelled.

The increased regulatory environment led to a significant rise in ballooning cancellation days during 2018–19; the now highly regulated nature of the balloons had significant ramifications on other aspects of ballooning and tourism in Göreme. For example, the frequent cancellations not only prevented many tourists from being able to fly – especially if they had come to Göreme for only a night or two – but there was also frequent disruption to activities, and businesses, surrounding the *Instagram balloons*. As mentioned above, also, the frequent cancellations sometimes led ticket prices to skyrocket, and this intense competition for tickets led to further concerns about tourists feeling "ripped-off" while in Göreme. On the other hand, the high number of weather-related cancellations meant that some *nuisance* aspects of the balloons were alleviated. Indeed, a decade earlier when I had asked a Göremeli businessman what he wished for with regard to Göreme's future, he told me that he wished for there to be "No balloons"; although the opposite had occurred with the growth in balloon numbers overall, the cancellations meant that, on some days at least, his wish was fulfilled.

Unintended consequences

By considering various characters or configurations of balloons as a multiplicity of actors coming into being at different times and some then dissipating, we can see how – along with the combined choreography of all the other 'dancers' (Franklin, 2012) – the balloons have enacted different tourism orderings over (and under) time. Starting out in the late 1990s as predominantly *Elite balloons* – as some sort of 'divine hero from the skies', the balloons had become *money-makers* a decade later, enacting a pervasive commission-payment system which in turn unleashed a rapid proliferation of hotel conversions, in turn prompting processes of resident displacement from the old neighbourhoods. During the following decade, in combination with another key actor – that of social media – the balloons took on a whole new 'character', the *Instagram balloons*. The *Instagram balloons* not only enacted Göreme's incorporation into the newly emerged globalised 'like economy', but they were also a key reason for "everyone turning to Göreme to do business". Indeed, Mustafa from the Göreme Tourism Development Cooperative continued our conversation referred to at the start of this chapter by telling me that, although promotion of Göreme for tourism used to be one of the Cooperative's main purposes, "there is no need for *reklam* (advertising) anymore"; "If anything", he continued, "we need to do the opposite, because it's all happening as a big force"; "It's all to do with the balloons", he repeated.

It is precisely because such a 'complex multiplicity of actors emerges … that tourism is [shown to be] never "pure" in its categorizations, never coherent in its planning and never truly controllable' (Jóhannesson, Ren and van der Duim, 2012, p. 166). In other words, it is the complex multiplicity of actors which has rendered tourism in Göreme seemingly out of control, or happening as a big force as Mustafa put it. Franklin (2012) sees this as 'an ontology of unintended consequences' (p. 46), in that tourism orderings enacted by *any* actors are:

> only ever *unleashed…* They may have blueprint beginnings, but as orderings that persist in time they have a more unbounded and open-ended

nature: they may not be confined to their intended object, they may not continue in the form initially conceived and they may have a range of effects, intended and otherwise.

(ibid., p. 46)

Hence, it is because the multiplicity of actors at play – human and non-human, balloon and non-balloon – have *only ever unleashed* potential tourism orderings, that they have together, over two decades, led to tourism in Göreme happening as a seemingly uncontrollable "force". It is thus in the point that this is an ontology of unintended consequences that we might ultimately view the balloons themselves *as the spectre*; the spectre of unlimited change.

Notes

1 Mehmet featured in *Living with Tourism (1)* in relation to his tourism 'Fred Flintstone' persona (see Tucker, 2003, pp. 170–172).
2 The 'like economy' is a term coined by Gerlitz and Helmond (2013) to discuss how 'like' and 'share' buttons on social media platforms, such as Facebook and Instagram, turn user engagement and feedback into numbers which can be traded and multiplied.

4 In search of the perfect self(ie)
Making stories, narrating self

> ...stories, as forms of expression, do not mirror the experience of the hunt. But if one measure of the success of the hunt is the story that is subsequently told about it, then what happens on the hunt is partly determined by *cultural notions of what makes a good story.*
>
> (Bruner, 1986, p. 17, *my emphasis*)

Referring to Ilongot hunt stories the above words from Edward Bruner's earlier writing made a crucial point about the relationship between story and experience, a point he later brought into his work on tourists and tourism. As Harrison (2019) noted: 'One only need substitute the word "trip" for "hunt" to capture Bruner's own insights about tourist stories' (p. 81). In making this link to tourist stories, we are prompted to think about how it is the tail (tale?) which wags the dog. In other words, what happens during travel is, at least in part, determined by tourists' *cultural notions of what makes a good story.* In reference to this point, Harrison (2019) reflects on her own travels:

> the bone-deep nauseating weariness as we rattled along on the wooden benches of yet another overnight train in Southeast Asia; or the intense panic and fearfulness I felt to my core when confronted with the masses of humanity in the streets of Mexico City. Just why did I choose to endure these things? I was never quite sure, but the emotional architecture of these experiences imbued them with endless capacity to fuel the stories I later told about my travels.
>
> (p. 72)

Tim Edensor (1998), too, pointed out that as stories are both 'an essential way of making sense of the world and transmitting identity' (p. 69), tourist narration inevitably involves pre-existing narrative conventions or, in other words, what makes a 'good story'. At the same time, however, 'the production of stories is also an ongoing process which may incorporate new elements and material' (ibid., p. 69). Such incorporation of new elements and material might indeed be how ideas around 'what makes a good travel story' change

DOI: 10.4324/9781003011200-4

over time. The focus in this chapter is how what makes a good tourist story in relation to Göreme has changed over the four decades of tourism there.

Besides cultural notions of what makes a good story, stories and the experiences they depict are, of course, contingent on certain affordances of place and other materialities which enable the good stories to be created and conveyed. Bringing Gibson's (1979) earlier concept of affordances into the context of tourism prompts a consideration of the ways in which different tourism places and materialities afford particular tourist behaviours and performances (Bærenholdt et al., 2004; Edensor, 2006; Kimber et al., 2019). In other words, the concept of affordances recognises the significance of how physical environments, as well as material culture and objects (such as hot-air balloons, Chapter 3), affect and act in the production and consumption of tourism. Germann Molz and Paris (2015) add to the discussion on affordances by looking at how new technologies – and particularly mobile devices and social media platforms – afford particular ways for contemporary travellers to convey their travel stories to a broad social network. Indeed, smartphones and digital connectivity have profoundly changed tourists' storytelling affordances.

In relation to Göreme, the changes in what makes and what affords a good tourist story also inevitably encompass changes in who the particular tourists visiting Göreme are, plus broader global trends in tourism. While the tourists who stay in Göreme's hotels and hostels have continued to be predominantly those travelling independently of a tour group, the changing tourist profiles and trends have inevitably constituted change in cultural notions of what makes for good tourist stories. Furthermore, over time, global trends in digital technology, which we might see as a key mode of storytelling, have interacted with the changing affordances in Göreme – in particular those pertaining to the growth in hot-air ballooning – leading to significant changes in the profiles and geographic origins of tourists visiting Göreme. The predominance of Western backpackers during the 1980s–90s became largely replaced in the 2000s, not only by more mature and more moneyed tourists from Western countries, but also by tourists coming from other parts of the world, including the Middle East, East Asia, eastern Europe, and Russia, and surrounding countries. In addition, a marked increase in Turkish visitors occurred. During my visits to Göreme in 2018–19, I met tourists from various countries, including Poland, Uzbekistan, Thailand, Moldova, Ukraine, Indonesia, the Philippines, Argentina, Columbia, Iran, Sri Lanka, and so the list goes on.

In *Living with Tourism (1)*, the chapter which focused on the tourists in Göreme was entitled 'The tourists: in search of serendipity', describing how the tourists in the 1980s–90s – then predominantly Western backpackers – sought adventurous and 'authentic' experiences; those were the experiences they deemed would make for *good stories*. While serendipitous and adventurous experiences continued to provide the material for good tourist stories for some, stories of connection came to the fore as the tourist profile changed during the 2000s. The advent of mobile digital technologies and the launch of major social media platforms then prompted further major changes. The title of this chapter signals what next became the key focus of tourists' stories – their perfect self(ie). As Taylor (2020) remarked: 'For better or worse, the travel industry has been transformed by customers' drive for the perfect selfie' (p. 64). My particular interest here is consideration of the part played by these changes in the developing spectre of unlimited change.

Stories of serendipity

> People go to tourist sites loaded with expectations and knowledge about what they will find there. They do, quite often, see what they went to see. But…they also find something else,… something they find through accident. It is the accidents and unplanned events that provide the basis of stories….
>
> (Neumann, 1988, p. 27)

During the 1980s and 1990s, serendipity-seeking travellers were plentiful in and around Göreme and I wrote in *Living with Tourism (1)* about 'an alternative tourist discourse that promotes the ideas of discovery and adventure' (Tucker, 2003, p. 39) being prominent in travel guidebooks' chapters on Cappadocia. This was in line with Elsrud's (2001) similarly identifying that the narrative of risk and adventure running through backpackers' travel stories was an integral part of the identity-formation for these 'anti-tourists'. At that time, Göreme was deemed by such travellers to be suitably 'untouristy'; not only did staying in Göreme spatially separate the backpackers from 'mass' package group tourists who were accommodated in larger hotels situated in towns outside of the Göreme National Park, but Göreme's small locally owned and run accommodation establishments and restaurants were attractive because of their *ad hoc*, semi-formal and vernacular character which afforded these mostly young backpackers the ability to assert an alternative and adventurous identity. Tourists I met during fieldwork in the late 1990s spoke about their preference "to be free [rather than] to being guided and controlled", liking the ability "to discover it all for yourself", and wanting to differentiate themselves from "typical tourists" who "don't have proper interaction with the local community".

I also wrote in *Living with Tourism (1)* about backpackers seeking 'a certain level of surprise events, or happenstance, in order to individuate their particular experience' (Tucker, 2003, p. 55), and hence to individuate their stories. Referring to serendipity as 'the art of making an unsought finding', Grit (2014) has argued that serendipities 'do not just happen by luck, but, for their becoming, require a keen eye and a responsive attitude' (p. 133). The earlier backpackers in Göreme appeared to possess this keen eye and responsive attitude, as they were generally 'open to unplanned and unpredictable happenings, open to possibilities' (Tucker, 2003, pp. 55–56). An American backpacker I met during fieldwork in the 1990s told me, for example:

One thing I like best about travelling is getting up in the morning and having no idea who I'm going to meet that day, or what I'm going to experience. Some of [the experiences] may not be particularly pleasant either, but they are memorable.

(Tucker, 2003, p. 43)

We might reasonably read 'memorable' in this case as also meaning 'storyable'. Indeed, such stories of chance or accidental happenings and encounters appeared to lead the backpackers in Göreme during the 1980s and 1990s to travel to particular places and in particular ways in order to deliberately invite serendipitous experiences. Backpackers were also often referred to as 'budget travellers', and travelling as cheaply as possible – such as taking local buses or hitchhiking and ending up somewhere you had not intended to go – opened their travel experiences to adventure and happenstance; thereby providing good tales to tell afterwards. Such serendipity-seeking backpackers enjoyed exploring the Göreme valleys on foot, discovering churches not written about in the guidebooks, or encountering a farmer in his field and being given a handful of apricots or grapes. Renting a scooter to ride around the Cappadocia region was another way of opening oneself up to unplanned and unpredictable happenings. A New Zealander staying in Göreme in the late 1990s had done this and he told me how he enjoyed the chance to stop in villages "where there was all sorts of life going on" (Tucker, 2003, p. 59). He said he was invited to a wedding in one village, and was invited to join in with some Turkish dancing in another.

The way that Turkish hospitality was performed in accommodation and restaurants in Göreme during the 1980s and 1990s also was conducive to unpredictable happenings and encounters, hence affording the kinds of experiences the backpackers were seeking. The provision of hospitality frequently involved taking guests on trips to watch the sunset, holding an impromptu barbecue, and dancing into the night to Turkish *saz* music in the courtyard or the cave of a *pansiyon*. As I previously commented:

The services and interactions in the tourist realm of the village have not been perfectly set up and smoothed over, and so tourists can expect a rather tumultuous but friendly production of services. The expectancy that things will not go smoothly is all part of the adventure.

(Tucker, 2003, p. 126)

Relevant here is Grit's (2014) idea of 'hospity', a term he explains as 'experience within spaces of hospitality which is not defined yet; the host-guest relationship and interactions are not pre-given' (2014, p. 132). Grit refers to such spaces as 'healthy' hospitality spaces; 'healthy' because, in line with the oft-repeated saying by Göreme hosts to their tourist guests, in those first decades, that "Everything is possible in Turkey", they afford surprise happenings – and they afford the open and responsive attitude needed to value and enjoy those surprise happenings.

Importantly, as well as aligning with the backpackers' like for places and services 'to *speak back to them*, to surprise them, to challenge them' (Tucker, 2003, p. 67), the 'hospity' in the backpacker pansiyons and restaurants in Göreme in those earlier decades was also enjoyed by their Göremeli 'hosts'. For local men, the *pansiyons* were a kind of liminal space, or free zone, where they could be relatively free from the village way of life and where they could enjoy convivial relations with tourists. As one pansiyon owner told me during my fieldwork in the late 1990s:

> We grew up with backpackers, and it is wonderful. You can talk with them, you can learn a lot from them. The package tour people... don't have time. We cannot talk with them, spend time with them... Backpackers stay here for longer so you talk with them, have fun with them and you get to know their lives too.
>
> (Tucker, 2003, p. 134)

This appreciation for the conviviality in tourist encounters in the early years of Göreme's tourism appeared to be matched by a similar appreciation on the part of the backpackers themselves:

> We like it here more than the coast because the people here are very open and friendly... It's really different because the people are so open here, you can feel closer to them, you know, you don't really feel like a tourist here because you can have closer contact with the people.
>
> (German backpacker, quoted in Tucker, 2003, p. 134)

As the tourists visiting Göreme started to change in the 2000s and the numbers of serendipity-seeking backpackers started to lessen, such interest in 'contact with the people' continued and so it became stories (tales/tails) of encounter and connection which came to 'wag the dog' of experience.

Stories of connection

Extract from 2009 fieldnote diary: Today I met a family from Switzerland with two young boys. We met at the hotel swimming pool and their boys played with [my son]. The father had been to Göreme in 1984, when, as a backpacker on a budget, he had slept rough in some caves. He had returned with his family in 2008, and while Göreme had changed a lot since his 1984 visit, he said "It's still got its magic", and so they decided to come again in 2009. The father had gone out hiking in the valleys on his own early that morning. The family would hang out by the pool until it became cooler in the late afternoon, then they would all go for a walk. I bumped into them again later that evening on the main street as they were heading to a restaurant to have dinner.

Extract from 2009 fieldnote diary: At Sunset Viewpoint this evening I got talking to a family from New Zealand – the parents had been to Cappadocia twice, as backpackers, in the early and mid-1990s, and had loved it so much that they had named their daughter after one of the valleys in the region. They came this time to show Göreme to their daughter, now 13 years of age, and to take her to the valley she was named after.

By the late 2000s the predominance of budget backpacker tourists had dissipated to a large degree, and in their place came more moneyed grown-up versions of themselves. Many had visited Göreme as backpackers some years, or even decades, earlier, and were now revisiting as more mature travellers, some with their families. They spoke of having made a connection with Göreme's "magic", returning again and again, and making friends with Göremeli people. Like the German backpacker quoted above, they too enjoyed the feeling of *not* being "like a tourist here because you can have closer contact with the people." Usually staying in the old neighbourhoods, they tended to enjoy walking through the narrow, cobbled streets, sometimes having encounters with neighbourhood residents, such as women sitting on doorsteps. A woman from Boston I spoke with during fieldwork in 2009, travelling with her daughter and small grandchild, told me how the previous day she had been struggling

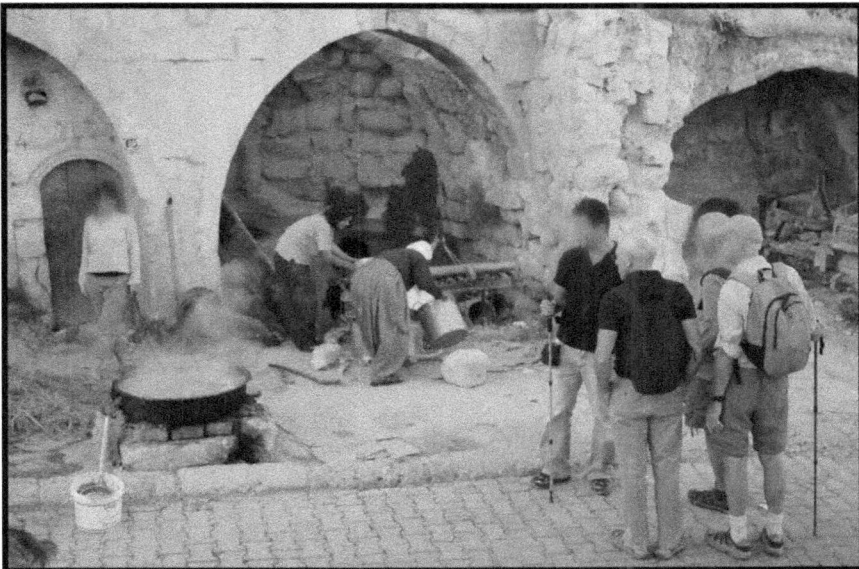

Figure 4.1 Tourists showing interest in local culture, 2009.

to push her grandchild's pushchair up the hill to their hotel when she passed a local woman, also with a small child, and they "exchanged a look of mutual understanding". The American woman told me she was enjoying hanging out in Göreme and "seeing how people live their lives". Similarly, an elderly couple staying in Göreme for a month told me they enjoyed "just hanging out" and going for walks every day, wandering in the backstreets of the village or having encounters with farmers while walking in the valleys.

In my fieldnotes from visits to Göreme during the mid-to-late 2000s, I noted tourists often telling me, or telling each other as they sat on their hotel terrace or around the pool, stories about meeting 'locals'. Stories of encounters and connection appeared to constitute *good stories*, in line with what Harrison (2003) describes as a general desire to experience human connection, sociability, and intimacy. Referring to Simmel's ([1910] 1971) 'sociability impulse', Harrison notes the desire for human connections as 'simply ends in themselves' (Harrison, 2003, p. 46) and argues that tourism encounters, from their outset, involve a 'willingness of both parties to engage in some form of connection' (ibid., p. 47). This point was conveyed to me by the Göreme Mayor during 2005 fieldwork. He said that, even though his elderly mother did not speak any foreign language, she was always warm towards everyone: "The other day", he told me, "she was carrying something heavy on her back and a "strong tourist" had taken it and carried it for her". She had come home afterwards and told her son (the Mayor), "My dear, they are really compassionate, these people". The Mayor went on to say that his mother always gives apples and grapes to tourists when they pass her in the fields, and he said that, because most of the people of Göreme do this, they have a good relationship with tourists and do not experience any problems with them. As Natalia Bloch (2021) has discussed in-depth, it is not only tourists who value making connections.

With tourism facilities and services becoming more formalised, however, the sense of connection afforded by Göreme's tourism dissipated somewhat over the years. In contrast to the informal nature and convivial character of tourism during the 1980s and 1990s which had afforded the building of significant relationships between tourists and Göremeli entrepreneurs, workers and other 'locals', the boutique hotels and other more upmarket facilities and services that developed in the 2000s tended to be more formal and customer-service-oriented. In general, the more upmarket a hotel became, usually developing a managerial structure in its workforce, the more formal and 'distanced' its atmosphere.

Nonetheless, the owners and the workers of even the more formalised boutique hotels frequently expressed a desire to get past what Bloch (2021) referred to as 'the dehumanizing nature of the service provider-customer arrangement' (p. 117). Indeed, Göremeli entrepreneurs continued into the 2000s to talk about how much they enjoyed getting to know those who returned and stayed in the same hotel repeatedly. With many of these guests being mature versions of their younger backpacker selves, they too continued to want to go beyond the service provider – customer framework and to be able to make a connection with their *Göremeli* hosts.

Other pockets of opportunity for 'connection' continued beyond the hotels and pansiyons. An example is 'Walking Mehmet', a self-created entrepreneur who started a walking-guiding business during the 2000s. As well as being promoted by hotelier friends around the neighbourhood in which he lived, Mehmet's walks were promoted in guidebooks such as *Lonely Planet*, as well as other travel literature such as *Wanderlust* travel magazine. Photographers and documentary-makers have often booked guided walks with Mehmet, resulting in international promotion. Incurring no particular capital expenditure, his guiding business provided Mehmet with a modest but ample living. Indeed, 'Walking Mehmet' was himself the tourism product, connecting with tourists via his physical agility and willingness to walk, as well as his extensive knowledge of walking tracks and hidden churches and other lesser-known points of interest in the valleys surrounding Göreme. Moreover, adept at helping those he guided to clamber through the cave-networks inside fairy chimneys, Mehmet enabled his guests to have a tactile and adventurous experience of Göreme's physical landscape.

Most of all, Mehmet provided close-up contact with an 'ordinary' local man (albeit with very good self-taught English language competency). Tourists would spend a whole day with him, or sometimes multiple days, chatting as they walked, connecting, and learning about how local people live and how they think. For instance, a Spanish woman, a doctor, had booked three days of walks with Walking Mehmet and, at the end of each day, she hung out with him all evening at his friend's café. Another "guest" Mehmet told me about was a Chinese engineer whom he took on a seven-hour walk, ending with sitting at Sunset Viewpoint drinking a couple of beers while watching the sunset and philosophising about life. Mehmet said that because the man had good English and was easy to talk to, Mehmet too had enjoyed their spending extra time together chatting about the 'rat-race' of life and the importance of striving for a calmer, happier life. Mehmet would undoubtedly have featured in both the Spanish doctor's and the Chinese engineer's 'stories of connection' from their time in Göreme.

Indeed, in Mehmet's own stories, he would always refer to those he guided as his "guests"; he told me that he refused to "use the word 'customers' because I am not selling a material thing – I am hosting". Many such "guests" developed a connection with Mehmet, returning and booking his walks time and time again. Some of Mehmet's repeat guests were international personnel posted in consulates and embassies in Ankara, who not only came to Göreme regularly themselves, but when they had guests visiting from their home country, they would bring them to Göreme and book to take them on a guided walk with Mehmet. Over the years, Mehmet therefore developed friendships with people in various 'elite' positions, providing him not only with a high level of intercultural competence but also a wealth of social capital which was likely to go well beyond the level of his monetary wealth.

Another example of a Göreme business persona who likely afforded tourists stories of connection over the years is Fatma, a middle-aged woman who, along with cooking for and part-managing her son's restaurant, set up cooking classes for tourists in her home's cave kitchen. Although Fatma's cooking classes featured on the *Lonely Planet* website, having limited English language ability, Fatma confided in me that she lacked confidence in her ability to communicate with tourists and answer their questions. Nonetheless, from my observations, for the tourists who learned to cook various Turkish dishes in Fatma's kitchen and then had the opportunity to sit with her to share the meal, the experience

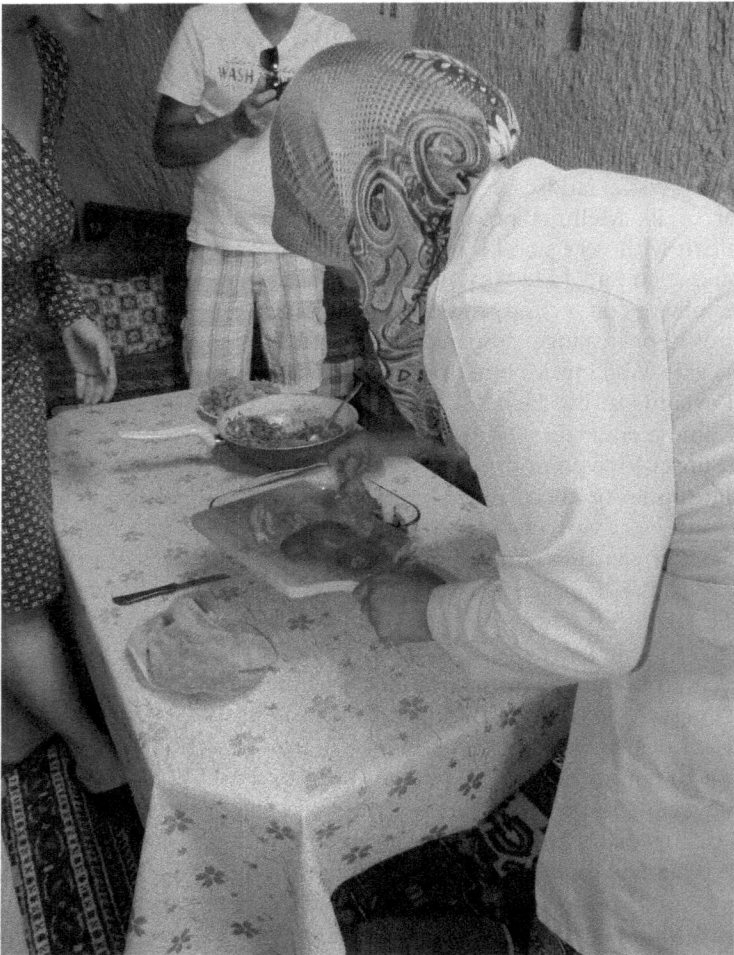

Figure 4.2 Fatma's Turkish cookery class.

was most certainly the kind of thing deemed to make for good stories of connection. A Dutch family enjoying a lesson with Fatma told me that they came to Turkey often as they visited Dutch friends who lived on the south coast, and then they would travel around. With their Turkey connection, they enjoyed buying foodstuff from a Turkish grocery store near their home, and so they liked the idea of doing one of Fatma's cooking classes so they could learn what to do with the grocery products they bought. The teenage children appeared to appreciate the class as much as their parents did and, sitting in Fatma's cave kitchen, they appeared to very much enjoy the meal after the lesson.

Other women, too, developed informal businesses by inviting tourists in to visit their cave-house, usually in the hope of selling them some self-made handicrafts. While these encounters appeared to be an opportunity for 'human connection' and thereby held much promise, many of the women – due to both their lack of English language ability and a lack of understanding regarding what tourists' expectations might be in such encounters – had difficulty in enacting smooth encounters with their tourist visitors. Any sense of human connection in such encounters therefore tended to be rather precarious and the encounters could often end up being awkward and uncomfortable for those involved (see Tucker, 2009a, 2011 for a more detailed discussion on this). Nonetheless, a few of the women did succeed in figuring out what worked and what did not work in their ways of interacting with tourists. An example was one woman who, selling self-made jewellery from her doorstep in the old neighbourhood, came to understand that her more zealous calling out to tourists inviting them to look at her cave-house and jewellery, as she had previously done, was perceived as too 'pushy' and was thus likely to scare tourists away. Over the years, this woman learned that a more reserved low-key approach to selling worked better; in more recent years, she would sit quietly and let tourists peruse her wares, giving them space to decide if they wanted to buy anything. If they did buy, and then if they passed by repeatedly in the days that followed, she might at some stage invite them to have a glass of tea, thus establishing a bit more of a connection and, no doubt, a story of connection for her guests.

While the business endeavours of Walking Mehmet, Fatma, and the doorstep handicraft-selling women were among only a few remaining examples in Göreme of tourism businesses which expressly afforded stories of connection, it could still be possible to gain stories of connection from other, non-business encounters. For example, I observed tourists having many such encounters with an elderly man named Mustafa (nicknamed 'Bicycle Mustafa' by Kaili and Lars for his coming by the Kapadokya Balloons office on his bicycle selling organic tomatoes from his gardens). In his retirement years, Mustafa always has time to chat and, with his white beard, round belly and traditional attire, Mustafa would often surprise passing tourists by breaking into conversation with them in competent English; he had learned English while working for a tourist shop before his retirement. Mustafa would often find a place to sit in the shade of a tree in the street by the central mosque and would get talking to tourists doing the same, or he would sit by the sweetcorn seller on the main street so that he could chat to tourists when they sat on the stools provided to eat their corn. One evening, in the summer of 2019, I bought a corn-cob and sat on a stool next to Mustafa. Soon afterwards, a young Italian couple bought corn and sat down next to us. Mustafa opened up a conversation with them and, through our convivial chat, Mustafa and I learned a little about the Italian couple and they in turn learned a little about us. After they left, another couple sat down with their corn, this time, the woman was from Turkey and the man was Dutch. Mustafa asked how they had met, and was interested to learn about their long-distance relationship of two-and-a-half years. The woman had been to Göreme before and wanted to bring her Dutch partner, but she commented how disappointed she was that Göreme had become all "*Instagram*-crazy" now. Both they and the Italian couple appeared to value their encounter with Mustafa and I by the sweetcorn stall, just as we enjoyed our encountering them.

Another typical interaction with tourists occurred when I saw Mustafa one Friday by the mosque after prayers and I came upon him sitting on a park bench talking with a young couple. I was helping the Argentinian couple take photos with Mustafa on their phones, when another elderly Göremeli man – an acquaintance of Mustafa's – came by and joined in the conversation. He then invited us all to go across the road to the teahouse for a glass of tea. There, another friend, a Belgian man who owns a small house in Göreme and lives there during the summer months, was phoned and asked to join us in the hope that with his having known Spanish, he would be able to help us communicate with the Argentinians whose English was limited. We proceeded to sit for quite a while, all enjoying conversation and having multiple glasses of tea. While the couple had visited some of the key sights such as the Göreme Open-Air Museum during their short stay, they became interested in going for a hike when the Belgian man and I described some of the valley walks around Göreme. They had not been for such a walk yet because they were not sure where to go; they were delighted that we were able to advise them and draw up a map to show them the way. However, they were scheduled to leave Göreme that evening on the night-bus to Istanbul and expressed regret that they had only just made all of these local connections when it was time for them to leave; "This was the best moment of our whole trip!", they said.

Not all encounters are so convivial, however. Even when, as is often the case, for the Göremeli person such an encounter is framed as hospitality first and foremost, for the tourists involved such informal encounters are often loaded with an array of ambiguities, particularly around the encounter's potential commercial elements, thus making tourists feel nervous and unsure. A brief encounter involving a middle-aged man from Brazil was one example. He had stopped to observe an elderly woman making *pekmez* (grape molasses) in the street outside her house, and when I saw him taking an interest I stopped also. Quickly ascertaining that I could speak Turkish and communicate with the woman, the Brazilian man asked some questions about the process of *pekmez*-making. Appreciating his interest, the woman offered him a taste of the fresh *pekmez*, spooning some out into a small bowl which she tried to hand to him. He declined and stepped back a bit. I encouraged him to taste it, saying how good it was, and so he accepted the bowl. When he then asked me "Should I give her a lira now?", it became apparent why he had initially declined the tasting; he had been nervous about taking something when he was unsure if he should pay for it or not. I gladly reassured him that, to the contrary, it would likely offend if he did try to give the woman some money, as her offer was purely a hospitable and friendly one. I have written elsewhere (Tucker, 2014) about another similar instance, but that time the tourists had given money – before I had a chance to stop them, leaving the *Göremeli* woman in that encounter extremely embarrassed. Nonetheless, while one or more parties in an encounter may at times get things wrong, it is perhaps the *intention* of connection that is important (see Tucker, 2014 for further discussion on this).

I saw this intention overtly in another encounter I was party to, this time when a young couple from Hong Kong were drawn into the house of a woman who was trying to sell bead jewellery and headscarves to passing tourists. The woman was my neighbour and I happened to be chatting with her as the Hong Kong couple came by; I was able to translate and help them converse with the Göreme woman. The young Hong Kong woman was a keen photographer and, enjoying taking close-up pictures of the Göreme woman doing her beadwork, she asked me to translate that her hands were beautiful. The Göreme woman asked me to suggest they buy a beaded headscarf, which I did. When the couple obliged and purchased a headscarf, the Göremeli woman then asked me to prompt them to buy more; perhaps a necklace or another headscarf. I attempted to 'soften' her wishes when I conveyed them to the Hong Kong couple, and they consequently did buy more, saying that the woman clearly needed the money. They looked around the house and asked me questions about the life in Göreme, and how life compares now to the presumed hard conditions of the past; they were delighted that I was able to tell them so much about it. I asked them what they had done so far while in Göreme, and they said they enjoyed going for walks, and seeing local life such as in this encounter. When they came to leave, the Hong Kong woman embraced the Göreme women to thank her and wish her farewell. When the Hong Kong man went to do the same, the Göreme woman blocked him and I explained that it was not proper for unrelated men and women to embrace. He apologised profusely and then, perhaps slightly ashamed about the botched embrace but otherwise seemingly happy, the couple went on their way, no doubt with a good story of connection to later tell.

The Turkish visitors who started to come to Göreme during the 2000s brought different meanings of 'connection' with them. Many were Turks residing in northwest Europe who came for a trip to Cappadocia while back in Turkey visiting relatives in their hometown. Cappadocia appeared to be for them a sort of playground where they could be free to reconnect with their homeland – their family and friends whom they now lived away from – in new ways; I observed many family groups enjoying clambering among the fairy chimneys, uncles reconnecting with nephews and nieces, and adult brothers re-enacting childhood competitiveness by racing to clamber high up on a rock-face. When I got chatting with them, they explained that part of the group was visiting from Germany or Holland and they had come to Cappadocia to "be together".

Some Turkish tourists – those residing in Europe and those from other parts of Turkey – appeared also to seek in Cappadocia a nostalgic connection with what life used to be like in Turkey. During the 2000s, a popular television soap opera set in the Ottoman era was filmed in various villages in Cappadocia, prompting Turkish visitors to come to the region not only to visit the villages and old houses where filming had taken place, but to visit the old neighbourhoods of villages such as Göreme where, they perceived, some of the "old ways" still continued. While this meant that Göreme people often felt "looked down-upon" as *köylü* (villager or peasant), creating a somewhat uneasy relationship between Göremeli and their Turkish visitors, this was by no means always the case and sometimes congenial connections were formed. For instance, I met a young woman and her mother from the south coast of Turkey while they were visiting Göreme in 2016. It was their first time travelling to Cappadocia and had decided to go there while – in the period following the attempted coup in Turkey – they could enjoy it without crowds of international tourists. The two women had made a connection with a man named Faruk who, normally a tour agency worker, had no work because of the lack of international tourists at that time. He enjoyed the women's company and, for no charge, he took them around some key sights, including the soap opera film-sets, and ate in restaurants with them in the evenings. Undoubtedly, the Göreme man became a central part of the two women's Cappadocia travel stories.

When Faruk had worked in the tour agency, prior to the 2016 attempted coup, he had enjoyed convivial interactions with tourists of many different nationalities; interaction which, too, would appear to have potential for a *good story*. He particularly enjoyed forming connections with tourists from China, theorising that it was because Turkey and China had a long-time connection because of the Silk Road that they were natural friends and got along well. He told me that Turkish people do not 'other' the Chinese like Europeans do, and so Chinese tourists get a better reception in Turkey. Indeed, Faruk appeared to create very good connections with the tour agency's Chinese customers, and a 'comments board' in the agency office contained many glowing messages in Chinese characters about his "deep humanity". Many of the Chinese tourists who came to Göreme, especially in the first few years after Turkey gained "approved destination" status with China and hence became a new destination for independent Chinese tourists, saw Turkey as an 'alternative' destination for young Chinese to travel to. Consequently, perhaps in a similar way to the European backpackers who came to Göreme in the 1980s–90s, these 'alternative' travellers likely sought experiences – and stories – through which they could confirm and perform their 'alternative' status; stories of connection with Göreme people such as Faruk were possibly among these.

As the 2010s proceeded, however, Göreme became so well-known that it was no longer an 'alternative' destination for anyone, including for Chinese tourists. A popular Chinese reality television show in 2015 portrayed celebrities staying in a cave-room in Göreme and taking a hot-air balloon ride, and that year, Göreme's popularity with Chinese tourists grew enormously. The biggest influencing factor that changed the way Göreme stood as a tourist destination, though, was the change in tourist use of digital communication technologies, and in particular, the increased usage of social media. Through mobile media and the Internet, tourists became more connected with elsewhere; not only their friends and family back home but a whole extended group of online 'friends'. As those connections with elsewhere became increasingly prioritised, connections with the *here* and now dissipated. Even in the earlier days of social media – the mid-to-late 2000s – I noted in my fieldnotes how many tourists appeared distracted from their 'real', or corpo*real*, context, spending their time on their hotel terrace with their laptop, busy on *Facebook*, or *Skyping* with family or work. Moving again into the next decade (2010–19), tourist use of mobile technology increased again and in my fieldnotes, I recorded hoteliers complaining that, now, their guests just sat around "on their phones all of the time", never looking around at Göreme itself. It was becoming apparent that stories of connection were no longer so sought-after. I asked Jessica Lee, travel writer and co-author of the Turkey *Lonely Planet* guidebook, whether she thought there were still, in 2019, tourists who seek encounters with local people. She answered:

> Yes, world-wide, but they're not in Cappadocia. There are loads of people who do want that, but they're not here… It is because of the way tourism has changed specifically in Cappadocia, and even more specifically in Göreme… Social media, and especially *Instagram*, has taken over, and it's made a massive change here to how people travel.

Göreme, it seems, had changed into something else for tourists; no longer was it a place known for affording stories of connection, but rather it had become a place for storying the perfect self(ie).

Storying the perfect self(ie)

"Göreme has become the No.1 spot in Turkey – all because of *Instagram*." (Kazakhstan tourist, 2019)

The launch and growth of many of the major social media platforms during the 2010s became heavily influential in travel and tourism during that decade.

Although photography and tourism have long been closely bound together (Crawshaw and Urry, 1997), the transformation that mobile technology and social media afforded was the new ability of travellers to post their photographs and convey their travel stories to a vast social network. In their writing about this transformation, Dinhopl and Gretzel (2016) suggested:

> While having photos of oneself is not new, the unprecedented ease and quantity of pictures of the self that tourists are able to obtain is new. What is also new is the ability to instantaneously share photos of the self with an online audience of known and unknown others.
>
> (p. 131)

As Jordan (2019) similarly put it, since story*telling* necessarily involves both creating the story *and* providing an audience with access to the story, digital platforms such as *Instagram* have offered a vast and growing affordance for tourist storytelling.

Furthermore, the entanglement, or merging, of travel photography, mobile phones and social media reconfigured 'what it is to "see" and "be" as a traveller in the world' (Kohn, 2018, p. 76) in that, when we travel, we are always entangled with 'our online social worlds where we are vying for attention and appreciation via social buttons in the "like economy"' (ibid., p. 73). As a resident foreigner in Göreme, Sara, told me from her observations of tourists, "These days we measure the success of our experiences by how much other people 'like' them, rather than by how much we enjoy them ourselves". Relatedly, Germann Molz and Paris (2015) have suggested that social media functions as a kind of 'statusphere', since it enables tourists to exchange 'stories' for status with a social network far beyond both their home network and their immediate fellow-travellers. Smith (2018) also notes that *Instagram* users specifically 'seek out locations that, once photographed and posted to their accounts, will accrue the most "likes"' (p. 183).

However, usually it is not the destination per se that is being 'posted' in order to accrue likes, but rather, as Dinhopl and Gretzel (2016) suggest: 'The destination remains important [only] in that it provides the environment that enables tourists to produce visual media of the self that is suitable for the story they wish to tell' (p. 136). Hence, the 'statusphere' and the tourist's quest to perform a particular view of their self – via posting the perfect selfie – have created a new dynamic between the traveller and the places travelled to. So while the statement that Göreme had become the No.1 spot in Turkey – all because of *Instagram* – might at first appear to suggest that tourists were propelled to see and photograph Göreme itself, it is rather that Göreme – the place – has become merely the backdrop for tourists to photograph themselves. As I wrote in Chapter 3, in the later years of the 2010s, Göreme and its surrounds became something akin to a gigantic outdoor photo studio, where tourists went in order to garner images of themselves which, once posted on social

Figure 4.3 A Göreme valley becomes a photo studio at sunrise.

media, would accumulate 'likes'. As Smith (2021) has commented, the new visual culture propels tourists to travel purposefully 'in search of the prime selfie backdrops that, upon posting, will boost their likes, follower count and ultimately social capital online' (p. 611).

I discussed in Chapter 3, using extracts from 2019 fieldnotes describing early-morning scenes on hotel roof terraces, at Sunset Viewpoint and in the valleys surrounding Göreme – how the hot-air balloons became a key affordance enabling '*like*'-able tourist stories for their social media postings. Whilst taking a balloon flight and being able to photograph oneself floating high above the Cappadocia landscape was one way of achieving '*like*'-able postings, it was the large number of balloons flying each morning that created a most impressive backdrop for the perfect self(ie). The steep ups and downs in the Göreme landscape, enabling tourists to easily access promontory vantage points, also contributed to Göreme's becoming akin to a photo studio for tourists. As well as particular promontory spots such as Sunset Viewpoint affording effective self(ie) photographs with the balloon-filled sky in the background, the topography of the village also meant that some hotels sat higher than others affording panoramic views from what became known as their roof-top '*Instagram* terraces'. Staying in a hotel with good panoramic views enabled an early-morning photoshoot without too much effort, although the popularity of some hotel's terraces meant that when the balloons were flying above each morning there were long queues of people waiting to get onto the terrace. I was told that one such hotel situated in the top part of Aydınlı

Figure 4.4 Tourists queuing on hotel roof terraces at sunrise.

Mahallesi had begun charging couples an extra one hundred euros to access its top-level terrace where an 'exclusive' photoshoot – without others around – could take place.

While the early morning when the balloons were flying became the most popular Instagram-photography time, Göreme and its surrounds were enacted as a photo studio at other times of the day as well. For example, Sunset View-point became very popular for selfie-taking as the sun set over Göreme and, without the balloons flying over, other photogenic props – such as camels and miniature horses for tourists to sit on to be photographed – would be brought up to the ridge in the early evenings. Other popular photography spots included the Göreme and Zelve open-air museum sites. Another area of fairy chimneys a few kilometres from Göreme known as *Paşabağı* became so popular for photoshoots that, in 2019, the Ministry of Culture and Tourism turned it into a 'protected' open-air museum by erecting a large fence around the area and charging an entrance fee. 'Wedding' photographs in these set-tings were popular particularly among Chinese tourists and Turkish couples, and it was not unusual while out in the valleys at any time of the day to see couples dressed in 'bride and groom' outfits, often with a hired photogra-pher and drone to take their photographs. This became a particularly common sight in what became known as 'Love Valley', formerly 'Honey Valley'. Within Göreme village also, and particularly around the old neighbourhoods, there were often groups of tourists, including 'wedding couples', walking around the narrow streets doing photoshoots with the old stone houses and doorways as their backdrop. An example was a group of three young Chinese women

Figure 4.5 Wedding couple and photographer in 'Love Valley'.

I encountered one evening while I was taking a walk around *Aydınlı* neighbourhood. I came across them in a small lane taking group selfies by the picturesque doorway of a cave-hotel; they had gone out – all wearing long dresses and wide-brimmed hats – specifically to do a selfie photoshoot in the orange glow of the early evening sun.

With the merging of photography, mobile devices, and social media, a new visual culture had developed which appeared to encompass a raft of new learned behaviours and 'implicit conventions of what these photos of the self can and should look like' (Dinhopl and Gretzel, 2016, p. 131). Indeed, the scenes I observed at the main *Instagram*-spots – on the hotel roof terraces, at Sunset Viewpoint and in the valleys around Göreme – were remarkable in the way that, as tourists attempted to story the perfect self(ie), the same photograph-related behaviours were repeated time and time again. On each occasion that I visited the Sunset Viewpoint ridge in the early mornings as the balloons were flying overhead, for example, among the hundreds of people eagerly running around trying to capture optimum photos of themselves, there were always several couples doing 'piggy-back-with-arms-out' poses, young women being photographed by their partner as they sat on the ridge doing 'the look back', couples and small groups of friends taking group selfies using selfie-sticks, and family groups having their picture taken while sitting on a Turkish rug or at a mock breakfast tray. Many people were wearing a costume of sorts, including women in long flowing dresses, various shapes and sizes of hats, and couples in bridal gown and wedding suit.

Figure 4.6 Tourist performing promontory witness pose near Sunset Viewpoint as the balloons fly by.

Especially common was what Smith (2018) has called a 'promontory witness' pose, where the photographed person would stand with their arms outstretched to the sides and one foot placed slightly behind the other with toe extended, 'looking' out at the balloon-filled sky. Smith (2018) argues that such a placing of self in the landscape 'draws on semiotic canons of the colonial picturesque and the Romantic sublime' and, as such, is 'deeply imbricated with discourses of power' (p. 1). Smith bases this argument around an in-depth analysis of tourists' use of the 'promontory witness' pose for *Instagram* photography, whereby the usually lone tourist is pictured sitting or standing at a vantage point gazing out over the otherwise 'empty' landscape. In and around Göreme, it appeared that several promontory spots afforded this particular photographic composition which works 'to demonstrate exclusivity and the rarefied status of the tourist' (Smith, 2021, p. 609); in other words, working to story the perfect self(ie).

The physical environment and other materialities in and around Göreme appeared to afford for tourists other Instagram-able stories and performances of the self too. For example, Turkish family groups – from nearby areas and those including Turkish migrants visiting 'home' from Northern Europe – appeared to use clambering around the fairy chimneys and rocky slopes to perform and photograph their 'playful' or 'adventurous' self. I was told by several young Chinese tourists that the cave-hotels and restaurants, as well as the 'Instagram terraces' decked out with Turkish carpets and cushions, created an aesthetic which afforded performances of a 'bohemian' self(ie). Also, several

Figure 4.7 Tourist couple with hired classic car and photographer.

groups of Chinese tourists told me that the hot-air balloons and the surrounding Göreme landscape afforded a 'romantic' story of self, and that a 'romantic' self-image is good – as in 'like-able' – for their *Instagram*, *WeChat*, and *Weibo* stories. Their 'romantic'-selfie performances were further enabled by hotels keeping a box of 'romantic' dresses, including wedding gowns for their guests to borrow for photoshoots and, on some hotel roof terraces as well as other promontory spots, erecting of large heart-shaped metal frames for couples to sit within to be photographed. Kimber et al. (2019, p. 178) have noted that while many young Chinese tourist performances on social media revolve around social capital and prosperity, 'for many, showing others that they are in love, is equally important' (Kimber et al., 2019, p. 178).

During the latter half of the 2010s, along with the hotels providing '*like-able*' settings – such as a roof terrace laid out with mock 'breakfast table' and Turkish carpets and cushions – for guest photography, other businesses in Göreme adapted, too, so as to enable tourists to photograph their perfect self(ie). For example, rather than continuing to focus on selling carpets, some carpet shops developed a hiring-out business so that tourists could hire carpets and 'traditional' costumes to take with them to Sunset Viewpoint or out to a valley to use as props in their photoshoot. One carpet -shop owner told me that a tourist from the Philippines had come into his shop and given him the idea to hire out carpets and costumes; "It was her idea originally and I was the first one to do it", he said, "but now other businesses have copied the idea so there is stiff competition these days." He went on to tell me that his shop had become famous with Thai tourists, "because they are big *Instagram* users";

"It's all about *Instagram* now", he said. Other, bigger carpet shops had designated specific rooms as photo-studios, charging tourists from ten euros for ten minutes of selfie-taking to 50 euros for a photoshoot with the shop's photographer and drone. The most *Instagram*-famous of the shops doing this had acquired very mixed reviews on *Tripadvisor*, with those leaving bad reviews reporting being shouted at by the shop's owner for attempting to take a selfie inside the shop before paying. Despite such a high concentration of negative reviews, it appeared that the pull of the shop's affording '*like*-able' self(ie) images meant that it continued to be popular with tourists nonetheless.

Other businesses adapted for the new visual tourist culture included tour agencies conducting '*Instagram* tours'. These tours would usually start off in the early morning, taking tourists for a photoshoot at balloon take-off sites first and then continue on to other well-known *Instagram*-able spots. A few agencies started offering private tours using classic cars and costumes as props plus a professional photographer added in to enable tourists to get optimum self(ie) images for their social media postings. Taxi drivers, too, could be hired to drive tourists around at 'balloon-time', enabling them to be in the right places at the right times to capture the perfect self(ie). One such taxi driver, who was from Nevşehir, told me that he used not to notice the landscape around Göreme; he "just saw it as rock". He said that these days, however, he had gained an appreciation of "how fantastic and unique the place is", and he especially enjoyed driving tourists around in the early mornings to help them get good selfies amongst the balloons. During our conversation, he showed me a collection of his own selfies saved on his phone, many of which included his taxi, and most had balloons in the background. This taxi driver, too, appeared to have been enculturated into the new 'perfect self(ie)' visual culture.

Tamara Kohn (2018) has posed the question of 'what happens when we add "the selfie" into the picture?'; 'What does it mean to turn our backs on nature and produce selfies to post to one's social network?' (p. 72). As if in response, Christou et al. (2020) refer to the 'attraction-shading effect' of the selfie, whereby the place or attraction becomes '"shaded" by the bodies and faces of the [tourists]' (p. 292). Putting this another way, Hartal notes contemporary tourists' blatant disregard for 'what is right in front of them' (2020, p. 15). This disregard was highlighted for me one day in the late summer of 2019 when I was helping elderly neighbours with their *pekmez*-making activities. One neighbour filled sacks with bunches of grapes so that I could stomp on them to get out the juice which would then be boiled for *pekmez*. The other woman stoked and added sticks to the fire so that the batch of juice which was already over the fire would boil more rapidly. Since we were working in a small side lane near to and clearly visible from the street running up through *Aydınlı* neighbourhood, I expected we would attract attention from the many tourists who pass along that street. However, while lots of tourists did pass by, their attention was directed instead to the colourful wall of a carpet shop across the street – colourful because the wall was draped with Turkish and Persian rugs in order to attract people into the shop. While I stomped on the grapes, I saw many tourists stopping for photos of themselves in front of the carpeted wall clearly deemed to make a perfect backdrop for a perfect self(ie). They were so thoroughly absorbed in doing this that they barely noticed the 'local' food-making activity going on nearby. Even if they did notice, they continued their selfie-taking in front of the carpeted wall unabated. As Dinhopl and Gretzel have summarised in relation to what contemporary tourists are seeking: 'Rather than fetishizing the extraordinary at the tourist destination, tourists seek to capture the extraordinary within themselves' (2016, p. 135).

Figure 4.8 A tourist has 'perfect self(ie)' photo taken by colourful wall, while disregarding 'local life' occurring nearby.

"That's all they care about…just these crazy photos"

One morning in the summer of 2019, I was out walking around the narrow back streets of Göreme when I saw a young couple by a low wall surrounding the yard of a house. They were using the wall as a 'table' on which to place their art materials as they worked on drawings of the chickens in the walled yard. I went over to look at their artwork and, telling them that I too do sketches of Göreme scenes, we got talking about my research in Göreme, and, in turn, they told me about their travels; they were from Poland, hitchhiking around, and slowly making their way to the east of Turkey. Later that day, I bumped into the couple again when I had hiked to the nearby village of Uçhisar. They had their backpacks on and said they were leaving Göreme because it was "too touristy"; they felt that Göreme was the wrong town for them. By the end of the 2010s, Göreme no longer afforded the stories – of adventure, of serendipitous encounters, the stories of connection – that the Polish couple were looking for. Instead, Göreme now provided innumerable backdrops and settings for the 'perfect self(ie)'.

Twenty years earlier, Göreme would have been exactly the right town for the couple; they too, after all, were ultimately seeking the extraordinary within themselves. Indeed, the focus on self is by no means new; the tourists' search for serendipity and the search for connection were also, in many ways, about crafting a desired narrative of self, be it an 'adventurous self', a 'cosmopolitan self', and so on. Well before the age of digital technologies and social media platforms, Neumann (1988) wrote the following about tourists' notions of what makes a *good story*:

> People take out the "bad" shots, and put the images in some order…an order that tells some story. They are stories…that suggest how we ambitiously devote attention to remembering ourselves, and presenting our experiences to others….
>
> (p. 26)

Undoubtedly, travel has always been in good part about tourists' self-performance and self-storying.

There was something particular about the new visual-cum-social media tourist culture, then, that engendered such a fundamental shift over two decades. This was likely Göreme's becoming absorbed into the 'statusphere'; as Göreme became a place *defined by* its '*like*-able' backdrops for storying the perfect self(ie), it somehow lost much of its conviviality as a tourism place or, to draw upon Grit's (2014) idea of 'hospity' again, Göreme was no longer a space of hospitality which was not-yet-defined. This major shift appeared to prompt a kind of nostalgia among Göremeli entrepreneurs; a longing for the way things used to be: "We can't have conversation and can't make a connection with these tourists", one long-term hotel owner told me. He went on to describe the conviviality that there used to be in Göreme's tourism during the

1990s: "great connection, conversation, drink beer together"; "Now we don't have that kind of contact with tourists", he said. Another hotelier similarly told me: "It used to be much closer between us and tourists. Now they have no culture, no taste. They're just interested in balloons and photographs. And they want a different kind of service from before". This sentiment was further illustrated to me one afternoon in the summer of 2019 when I was standing with Osman, the hotelier introduced in Chapter 2, on the terrace of his new hotel. In a shared mood of cynicism, we counted the number of heart-shaped '*Instagram* frames' we could see on the terraces of other hotels around. Osman said "That's all they care about now – they just come here to take photos of themselves on these terraces. They're crazy – they're not interested in culture, or landscape or people, just these crazy photos". When I asked Osman where the guests in his new hotel tended to come from, he said he did not know. His not knowing, or more pertinently perhaps, his not being *interested* in knowing where they were from, conveyed just how much the tourists visiting Göreme these days – in search of the perfect self(ie) and propelled by social media and a seemingly unlimited quest for 'likes' – were integral to the emerging spectre of unlimited change in Göreme.

5 Changing life in the mahalle

Neighbourhoods, gentrification, and displacement

For ...what does the mahalle become if it is no longer the space of the familiar?.

(Mills, 2007, p. 351)

Fieldnote diary, 2019: Today in the late afternoon, I joined three elderly women as they sat out on the steps of a small side street in the lower part of the neighbourhood. They told me how hard their life had been. Two of them had had no education at all and were illiterate, while one of the women had completed five years of schooling and could read and write.

One of the women – the one who had been to school – pointed over to a nearby roof-top restaurant and said her son had made the restaurant on the top floor of the family house. She and her son lived in the remaining rooms underneath the restaurant, and she joked that she was 'the boss' of the restaurant. Another woman said that she planned to sell her old cave-house and, having bought a plot of land in the Yeni (New) mahalle together with her grown-up sons, she would have a large house built for the extended family to live there; "It's nice and quiet down there", she said, "here it's become too noisy". They all agreed with this. I asked how life in the old mahalle had changed over the years. They said that on the one hand it had become very good because "everybody has bread-money now". On the other hand, they added, there were no neighbours left. They acknowledged that the three of them were lucky because at least they still had each other as neighbours in their (lower) part of the mahalle. They pointed up the hill towards the upper part of Aydınlı Mahallesi, and across to Gaferli Mahallesi, and said "It's all hotels up there and over there – there are no neighbours left!" "And it's too noisy with all the traffic", one of them added. The woman who was planning to sell then started telling the third women that she should do

DOI: 10.4324/9781003011200-5

the same, especially as her sons lived in rented accommodation: "If you sell your house here", she told her, "you will be able to buy apartments in Nevşehir or Avanos for all of your sons". The third woman replied that she had no desire to sell and move; "I know here, not there; I'm comfortable here, so I don't want to sell".

Mahalle is the Turkish word for neighbourhood and, according to Amy Mills (2007, p. 339), it is 'the Turkish cultural space of closeness and familiarity produced through practices of neighbouring that create bonds of "knowing" (*tanımak*)'. Neighbourhood is a word that has increasingly entered into discussions about tourism gentrification effects; in turn, gentrification is broadly considered to be 'an economic and political process with structural capacities, actors, and actions that have the power to determine the contours of many people's lives, the social character of the neighborhoods' (Özbay, 2023, p. 2). Gentrification effects in tourism contexts have generally included concerns about displacement and marginalisation of neighbourhood communities (Cocola-Grant, 2018; Gravari-Barbas and Guinand, 2017; Wilson and Tallon, 2012). While becoming a worldwide problem, commentators and protestors at the local level have increasingly blamed tourism for killing neighbourhoods; thereby placing the neighbourhood as tourism's 'murder victim' (Germann Molz, 2018). While such sentiments tend to be related to extreme cases of displacement of original inhabitants due to tourism-influenced increases in land and property values and rents (Neef, 2021), there are a variety of more diffuse ways in which 'tourism gentrification involves a deep mutation of the place in which long-term residents can lose the resources and references by which they define their everyday life' (Cocola-Gant, 2018, p. 289). Cocola-Gant (2018) terms this process 'place-based displacement' which, alongside actual departure or displacement of residents, describes the process whereby 'residents feel a sense of dispossession from the places they inhabit, or "loss of place"' (ibid., p. 288).

While, in a broad sense, this book tells the story of tourism-induced place-based displacement occurring in Göreme as a whole, in this chapter I will focus on specific Göreme neighbourhoods in order to consider how tourism gentrification's deep mutation of place is experienced in transformations of everyday life in the mahalle. Moreover, highly pertinent in the Turkish context are the 'multiple linkages between gender and space in everyday neighborhood life' (Mills, 2007, p. 336), and hence what Sakızlıoğlu (2018) refers to as the gender-gentrification nexus

Gentrification alters the ways in which places are gendered and in doing so it reflects and affects the way gender is constructed and experienced. The ways places are gendered, as well as changes in gender constructions, may also affect the occurrence of gentrification.

(p. 205)

Indeed, in *Living with Tourism (1)*, I described, 20 years ago, a gendered socio-spatial duality in Göreme, an awareness of which was 'present in the minds of villagers and tourists alike' (Tucker, 2003, p. 8):

> The tourism zone, in the village centre, is where tourists are serviced with fun and entertainment, and where new building and tourism business is permitted and plentiful... [whereas] up the winding residential streets away from the village centre,... tourists and tourism are not so visible and the village women get on with their daily routine of domestic activities.
>
> (Tucker, 2003, p. 8)

Delaney (1991), too, referred to a socio-geographical duality in the Turkish village she studied whereby the *aşağı* (lower) section of the village was considered 'modern' and *açik* (open), while the *yukarı* (upper) mahalle was seen as 'traditional' and *kapalı* (enclosed). This duality between açik (open) space and *kapalı* (enclosed) space is at the crux of why it is important to consider here the everyday gendered experiences of gentrification and displacement. In Göreme, where the residential mahalle was previously *kapalı* space, these days, the narrow winding streets of Göreme's mahalles have more hotels than residential houses, and in some mahalle streets, there are restaurants, tourist shops, travel agencies, and, more recently, bars. Not so visible today are neighbours and neighbourliness in these neighbourhoods.

It should be noted that tourism is not the first disruption of neighbourly life in Göreme, however. A sense of ambivalence has long surrounded the way of life in the old mahalles. During the 1960–70s, not only were people moving away to seek work and a more comfortable living in Europe or other parts of Turkey, but there also, across the Cappadocia region, was a government push to resettle people from their old cave-houses into new "European-style" housing. The main reason for this was the old cave-dwellings being considered unsafe or unfit to live in (Emge, 1992). In some areas, and particularly in Göreme's Gaferli Mahallesi area (the neighbourhood situated below the Sunset Viewpoint ridge and where the old cave-house of Abbas's family was located), the real risk of rock fall or collapse in and around houses prompted government construction of "*Afet evleri*" (translating as disaster-relief houses) in the lower section of the village. Besides this real danger of the often dilapidated cave-dwellings, a sentiment emanated both from authorities and the communities themselves that it was shameful to live in such "backward" conditions. In contrast, built in neat rows out of concrete blocks and with a red-tile roof, the *afet* houses were considered "modern" and, at the time, carried a significant amount of prestige. Hence, even where there was not an actual threat of collapse, in order that they could apply for one of the heavily subsidised *afet* houses in the Yeni (New) Mahallesi, villagers claimed that such a threat to their old house did exist and applied for an *afet* house. Consequently,

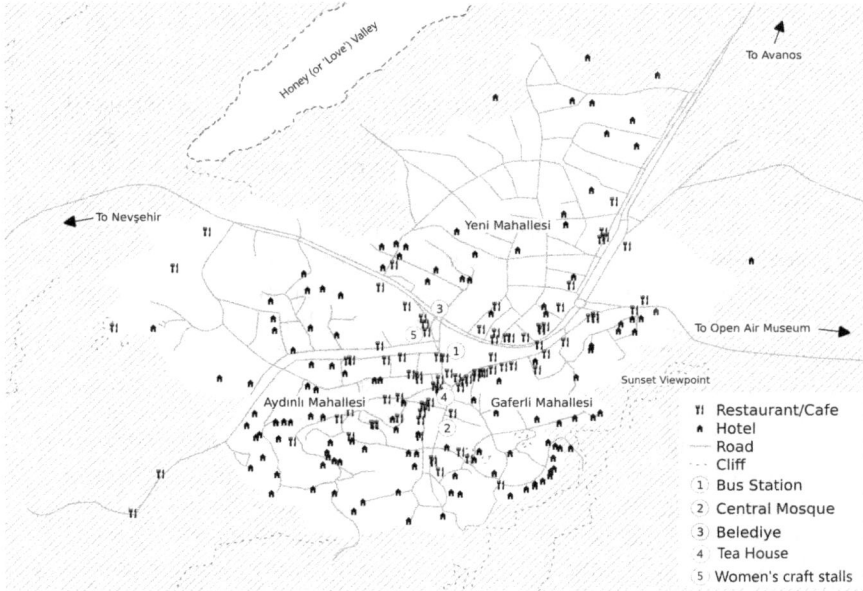

Figure 5.1 Göreme map showing neighbourhoods/*mahalles*.

whole sections of the old mahalles, particularly in the upper parts of Gaferli Mahallesi and Aydınlı Mahallesi, were abandoned. While there was a reduction in neighbours and neighbourliness in the old mahalles at that time, in their place, new neighbourly relations were formed in the streets of the new mahalle. Many families also maintained links with their former neighbours, continuing to attend weddings and other festivities in their old mahalle, and some households continued to use their old cave-house as a space for food production and storage.

In this chapter, while I will refer mainly to three of Göreme's mahalle, Aydınlı Mahallesi, Gaferli Mahallesi, and Yeni (New) Mahallesi, I will dwell on processes of place-based displacement in the old neighbourhood of Aydınlı Mahallesi in particular. I have developed 'neighbour' status, at least in part, in Aydınlı Mahallesi as that is where I have tended to stay – and more recently where I have my own cave-house to live in – when I go to Göreme. It is there that over the years I have created the strongest neighbourly 'bonds of knowing' (Mills, 2007, p. 339). According to Mills, this 'idea of 'knowing' (*tanımak*)' is what 'defines the mahalle: everyone 'knows' each other, or is 'known' in the neighbourhood' (ibid., p. 341). From my following of the changes to neighbourhood life over time, I see the loss of 'knowing' and familiarity as being at the heart of the place-based displacement ensuing from tourism in Göreme; hence, at the heart of the spectre of unlimited change.

Houses and (gendered) neighbourliness in the mahalle

Extract from *Living with Tourism (1)*: 'I met Hava because I used to pass her house regularly and she would be sitting on the doorstep of her own or neighbouring houses together with her friends, neighbours and their mothers. These women and girls would usually be busy crocheting or knitting, or preparing vegetables or fruit, but also sitting where they could observe passers-by, some of them tourists. Göreme women's lives are strictly governed by codes of shame and honour, and are very much centred around the home and the immediate neighbourhood... From spring to autumn Hava spends much of the day working with her mother in the family gardens in the valleys surrounding Göreme. Depending on the season, this work may involve tending and harvesting grapes, wheat, or apricots as well as gathering grass for the cows and donkey and the gradual collection of wood-fuel for the winter. In the afternoon they come back home and, after doing any cleaning and food preparation needed that day, get together with friends and neighbouring women to sit and chat'.

(Tucker, 2003, p. 72)

This extract from *Living with Tourism (1)* provides a memory of my earlier fieldwork visits around the mahalle, when Hava[1] was a teenager living in her family's cave-house in the centre of Aydınlı Mahallesi. I became close friends with Hava and her group of neighbours during my fieldwork in the mid-1990s, and would often spend time with them in the mahalle, sitting on doorsteps chatting and watching passers-by, and frequently I joined Hava and her mother when they went, using their donkey as transport, out of the mahalle to their gardens. Situated right in the middle of Aydınlı Mahallesi, Hava's family's house was typical of the three to four-hundred year old cave-dwellings in the villages of Cappadocia. Built around a central courtyard, or *avlu*, the rooms were a combination of cave-rooms dug out of the tufa rock and rooms built out of stone blocks which would have a flat roof suitable for laying out fruit to sun-dry. Also typical was that the house was set into, and up, a steep hillside or cliff, with the courtyard at the bottom, surrounded by a high wall (for privacy of the women inside) and entered through a large wooden door from the street. Together with the donkey, two cows were kept in the cave-stable in the lower part of Hava's family's house.

As part of my fieldwork in the 1990s, I gave a small camera to Hava and asked her to use it to record the main aspects of her life. It came as no surprise that all of the photographs she took were taken either in or close to her home, and that over half of the photos were of her female friends and rela-tives. Furthermore, whilst some photographs showed Hava and her friends relaxing, there were more depicting work and food-production activities, such as her mother kneading bread dough and a group of six women neighbours communally baking bread. Some of the photographs showed panoramic views of Göreme, roof-tops, and fairy chimneys. As reported in *Living with Tour-ism (1)*, I read these photographs as Hava expressing her sense of belonging to Göreme (being '*Göremeli*'). However, as the pictures were all taken from around Hava's house, it may have been more accurate to read them as express-ing her sense of belonging to the mahalle. As Mills (2007) contends in her study of gender and mahalle (neighbourhood) space, the mahalle is the pre-dominant space of belonging in Turkish culture.

Although Mills' study was based in an urban neighbourhood of Istanbul, her analysis and descriptions of mahalle life, as well as the links she draws between the mahalle's spaces of daily life and the production of gender, are highly relevant and in many ways mirror my earlier descriptions of mahalle life in Göreme. I wrote in *Living with Tourism (1)*:

> For women, the mahalle is where they spend almost all of their time, and their female neighbours are therefore those with whom they form their closest relationships. Only if they have relatives living in another mahalle would women visit another quarter [mahalle]…A woman would never, except perhaps occasionally for a wedding, go to quarters of the village where she does not have any relatives. Within their own neighbourhood, on the other hand, women may move around quite freely. They often visit each other's houses, and groups of women often sit in the after-noons either on their doorsteps or outside their houses in the wider more open parts of the mostly narrow streets.
>
> (Tucker, 2003, pp. 79–80)

That women could move around the mahalle quite freely is closely linked to the idea, mentioned above, of the mahalle being *kapalı* (enclosed) space. As will be discussed in more detail in the next chapter, being *kapalı* has strong associations of propriety for women in Turkish village society, with *kapalı* de-noting covering of the hair with a headscarf, appropriate clothing to cover the body, and staying 'enclosed' within the household and neighbourhood. The amount of head cover the Göreme women would wear in different spaces indicates the *kapalı* associations of the mahalle, as explained in the following extract from *Living with Tourism (1)*:

> When inside the home, during the day when no men are around, women wear the small head-scarf (*yemeni*) typical of the region. Outside of the home, but within the immediate neighbourhood, they still only wear this small head-scarf, although more securely than in the house. When they

leave their immediate neighbourhood, they secure the small scarf so that it covers their chin and mouth, and they put on a second larger scarf (çarşaf) which drapes down to cover the full top half of their body... All women wear the outer scarf if they pass through the centre of the village or visit the market place.

(Tucker, 2003, p. 80)

In line with Mills' (2007, 336) noting that the mahalle is 'a space which extends the interior space of the family to the residential street', in Göreme's mahalles there was not a defined boundary between the inside of the house and the outside streets of the mahalle. The binary of public and private is thus unable to describe the workings of space in the mahalle; women were not 'hidden' away inside their homes because, as much as their house, the mahalle streets were *their* space. Frequently sitting chatting on a doorstep, on a wall, or on the side of the street itself, the mahalle was a place where women could comfortably be outside, both in the company of others and in view of others.

The actions of daily life that linked neighbours together also included frequent popping in and out of each other's houses participating in *komşuluk* (neighbouring); even today, houses in Göreme's mahalles remain unlocked during the day when the women of the household are at home, with the door or gate often left ajar so that neighbours know they can simply come in:

Visiting is a natural part of daily life, making the inside sitting or visiting space of the home always open to visitors, extending this interior space onto the street... These visits create the 'knowing' of everyone by everyone else, a primary defining quality of the Turkish mahalle.

(Mills, 2007, p. 343)

Such practices of *komşuluk* (neighbouring) and *tanımak* (knowing) were thus precisely what produced the mahalle as a space of familiarity and belonging (Mills, 2007).

Komşuluk (neighbouring) practices and the fluidity of movement between the interior of the house and the mahalle streets were also precisely what tended to be missed terribly by those who had moved away, or had been displaced, and had gone to live in town. As Erman (2016) suggested in her discussion of displacement from Ankara's *gecekondu* (slum housing), 'the practices of sitting in front of their houses, baking bread in their gardens, etc., which constructed the fluid boundaries of private and public in *gecekondu* areas' (p. 213) were precisely the *means of social reproduction* women were dispossessed of, and missed, when relocated into social housing estates on the edge of the city. Likewise, when women (were) relocated from Göreme and moved to Nevşehir where housing in the many new apartment blocks was much more affordable than Göreme's land and housing, they tended to become somewhat confined within the walls of their apartment. What they sorely missed was the daily life in Göreme mahalles where they were not confined

Figure 5.2 Me sitting (third from left) with neighbourhood women practising *komşuluk* during 1997 fieldwork.

within the house, and where they were able, on a daily basis, to participate in *komşuluk* (neighbouring).

Whilst *komşuluk* (neighbouring) practices pertained largely to the 'traditional gender roles of wives and mothers which place them at home during the day' (Mills, 2007, p. 336; see Chapter 6 in this book also), the neighbourly bonds created among boys and men were strong and important also. As mentioned in Chapter 2 in relation to Abbas and Osman who were childhood neighbours in the Gaferli mahalle, for example, many friendships as well as successful business partnerships among adult Göremeli men derived from their having grown up in the same mahalle. As Mill's (2007) notes, the 'mahalle as a social collectivity is likened to the extended family' (p. 340), with the 'actions of daily life that link neighbours together in bonds of sharing, support, and reciprocity' (ibid.). Stirling (1965, p. 149), too, earlier pointed out that in Turkish village life 'no hard line exists between kinship – *akrabalık* – and neighbourliness – *komşuluk*'.

Just as groups of women sitting outside together in the street extended the interior space of the family home outside to the street and the mahalle, so too did the point that households would often conduct their work outside in the mahalle streets. This generally happened in autumn, when food-production work would often involve lighting a fire to cook on. When the courtyard of a house did not have suitable space to light a fire, the street became an extension of the courtyard, and neighbours would gather around, help out for a while, and perhaps bring an earthenware pot filled with the ingredients for a bean or meat stew to cook in the fire for their evening meal (see Chapter 7 for further discussion of these practices). These days, it tends only to happen in the Yeni (New) Mahallesi that family or neighbour groups might block the street in order to make room for a fire and the associated food-production work; tourism has now encroached so much in the old mahalles that remaining residents generally feel unable to disrupt the tourism-related traffic in order to undertake household activities. In any case, these days there are too few neighbours remaining in the old mahalles to help with such work.

Early tourism in the mahalles

Village life in the old mahalles has long attracted interest from international visitors, just as it did from the group of American national park 'expert' advisers who advised on the creation of the Göreme National Park. In the American group's report,[2] it was recommended that because the picturesque "activities of the villagers" added a resource to the general landscape of the park, the villages – and the villagers – should be allowed to remain within the Göreme National Park when it was formed in 1985. Upon recommendations made in the report, Göreme village – which was the largest settlement within the park boundary – was divided into various 'preservation' zones and 'tourism', or business, zones. New building was only allowed to occur in the central area of the main street, as well as in the Yeni (New) Mahalle at the lower end of the village. In the old mahalles, which were zoned for 'preservation', new buildings were not allowed and any alterations or restoration work to old houses or fairy chimneys required permission from the *Koruma* (Protection) office in Nevşehir, ostensibly 'protecting' their alignment with the aesthetic of the original structure. As well as protecting the fairy chimneys dotted around the old mahalles (some of which 'house' 1,000 year old Byzantine churches), this '*koruma*'/'protection' was intended, initially at least, to preserve the old houses (many of which are 300–400 years old) and the "picturesque village life" of the old mahalles.

However, this valuing and 'protection' of the "picturesque village life" within the auspices of the Göreme National Park created a context of contradiction and ambivalence, since village life – and especially that within cave-houses – was simultaneously viewed as "backward" and even "shameful" in other regards, such as by national government and regional authorities. The government's '*afet*' (disaster-relief) programme mentioned earlier, which re-housed families from the old cave-house mahalles into newly built houses in the Yeni (New) Mahallesi during the 1960s and 1970s was, at least in part, a manifestation of this negative valuing of the old neighbourhoods. It was thus under the two conflicting programmes – the '*afet*' re-housing programme and the national park formation programme – that there developed something of an ambivalence amongst the villagers and the authorities regarding the valuing, protection, and preservation of the cave-houses and cave-life in the old mahalles. In the midst of this ambivalence, there also emerged awareness of a growing valuing of the old mahalles' cave-houses and cave-life by and for tourism.

Seeing tourists' interest in the caves and the cave-life, during the early years of tourism in the 1980s, villagers became aware of the aesthetic value and economic potential of their cave-houses. Some of the older crumbling cave-houses that were previously abandoned when households moved into '*afet*' houses in the Yeni (New) Mahallesi were consequently restored and turned into cave-house pansiyons by the original owner. A few households that had remained in their old house established 'family pansiyon' accommodation, of-fering accommodation in rooms of their cave-house and with the whole family involved in the provision of hospitality to tourists. As these pansiyons became busier, however, the family unit tended to be moved into another house so that the original cave-home could be turned completely over to tourist accom-modation. By the mid-1990s, there were approximately 50 cave-house pan-siyons in the village (Tucker, 2003), and this number grew to approximately 60 by the mid-2000s. Dotted around the old mahalles, these cave-pansiyons brought tourism into and amongst the life of the mahalle.

As well as being suitably vernacular in appearance, the pansiyons gave tour-ists a strong sense of being hosted by the Göremeli owner. Many owners made a point in their advertisements of pansiyon being set in what was the owner's original family home. The following is an example from the website of Kelebek Pension, in 1998:

> Kelebek Pension has a wonderful location, right in the heart of Göreme's historic old village, with spectacular views looking out towards Uçhisar and Çavuşin. Until 1993 Kelebek was my family home, and like most homes in Göreme, it was made of a mixture of cave rooms and stone-arched rooms which had been added later. Today the pension has 16 rooms, some of them in the fairy chimneys, some in caves and some of them in the traditional arched rooms where my family used to live... We hope you will enjoy trying out the cave lifestyle with us...

In those early years of pansiyon development, the Göremeli owners provided a fairly informal experience of "Turkish hospitality" for their tourist guests; serv-ing meals, taking tourists on trips in their cars, and drinking beer together and singing Turkish songs when the tourists gathered in the evenings.

Sometimes their hosting also included performing being 'troglodytes' in a cave-land fantasy (Tucker, 2001, 2002). Names of pansiyons included 'Flint-stones Pansiyon', 'Troglodyte Pansiyon', and 'Rock Valley Pansiyon', creating a sense of a fantasy-land where tourists themselves could stay in a cave. The pansiyons' separation from everyday mahalle life was aided by the way in which the original village houses had been constructed, with the rooms being set around a courtyard which was separated from the street by a high wall. The wall meant that pansiyons were set up so that tourists could see out, by hav-ing access to upper-level roof terraces in order that they could enjoy the view of the village life below and the valleys beyond, but people outside of the pansiyon could not see in. I have earlier written how these accommodation establishments thus acted as a sort of liminal zone for Göreme men (see Tucker, 2009b), where they could be relatively free from the usual rules of village life; where they could drink beer, wear "tourist clothes", such as shorts, meet (and sometimes sleep with) foreign women, and generally hide from the watchful eye of their elders. Although set in villagers' old family homes and mixed in amongst mahalle life, the pansiyons acted like small islands of tourism which were in amongst, but also separate from, daily mahalle life; hence, the mahalles continued to be more or less *kapalı* (enclosed) space even though tourism was developing within them.

Outside of the pansiyons, another form of gendered tourism business emerged, this time involving some neighbourhood women capitalising on their cave-houses by inviting tourists in to view their house and then attempting to sell them handicrafts. As discussed in Chapter 4, tourists walking to and from their accommodation, or wandering simply to explore the back-streets of the village, often encountered the neighbourhood women going about their daily life. Sometimes, they would experience an encounter such as the one I described in an earlier discussion of the workings of these encounters (Tucker, 2011) as follows:

> Emine asked the tourist couple if they wanted to see her cave house. They said they did and I went too, at the request of Emine, in order to translate. Through Emine's front door we came out onto the terrace and the German couple marvelled at the view over the village and the Göreme valley. Emine then led us into a sort of make-shift cave kitchen and sat us down and offered us grapes, saying that they had come from her sister's garden... We then went into a beautiful arch-roofed room, the room normally used for receiving guests. In there, Emine immediately emptied the contents of a plastic bag onto the floor. She invited the German woman to sit down and have a look at the pile of headscarves and knitted socks now on the floor...
>
> (Tucker, 2011, pp. 29–30)

Otherwise largely separated from tourism business, this opportunity to engage with tourists within their mahalle space allowed women to engage with the tourism economy and to earn some money of their own. As I will discuss in Chapter 6, their entrepreneurial activities allowed these women to 'craft new selves', a term used by Cone (1995) in relation to informal female entrepreneurs in Chiapas, Mexico. Unlike the Chiapas women studied by Cone, however, the Göreme women did not '"step outside" their domestic spheres' (Cone, 1995, p. 315) in order to sell handicrafts to tourists; rather they brought tourism *into* their cave-homes. These entrepreneurial activities thus worked to reconfigure their home space, while also necessitating a negotiation of the ways that gender identities were performed within that space (Tucker, 2007, 2009b, 2011).

While these informal entrepreneurial activities enabled women in the old mahalles to 'craft new selves' to some degree at least, they also created tension among neighbours. There was much gossip around the mahalles about how the women's waiting on their doorstep for tourists to pass was akin to begging, and the owners of nearby pansiyons and hotels disliked the practice, saying that it made their hotel's guests feel uncomfortable when they were approached by the neighbourhood women and "pushed" into buying handicraft items. The owner of a hotel situated in the street beyond Emine's house referred to above, for example, said that he wished Emine and her neighbours would stop approaching *his* tourist guests in this way. Similarly, other women in the same mahalle – neighbours of the women trying to sell handicrafts to tourists – directed criticism towards these practices saying they gave a bad impression of the mahalle. An elderly woman who did not engage in this activity told me of those who did: "It's more than shameful. I could never do that. If you are going to work, go and work like a person". Such castigations not only conveyed the idea of the neighbourhood as a key space of propriety, but they also highlighted emerging tensions surrounding tourism in the mahalles.

Gentrification and loss of neighbours in the mahalle

Picking up a pace from the early 2000s onwards, increasing numbers of householders sold their old cave-house, mostly to keen hotelier buyers and sometimes to foreigners. At this time, '*Satılık*' (for sale) signs in both Turkish and English started appearing on walls and gates all over the old mahalles. These signs enabled visiting tourists to imagine owning their own cave-house there, and in turn, they allowed villagers who owned an old cave-house to imagine the large amount of money they might get if they sold to a *yabancı* (foreigner), or '*turist*', who tended to pay much higher prices than local people. In 2005, I recall standing on the terrace of a hotel looking out across the upper parts of Aydınlı Mahallesi and chatting with the hotel's owner as he pointed out the many houses now owned by foreigners: "That one belongs to Kaili and Lars (the hot-air balloonists discussed in Chapter 3), that one to Pat (at the time, the writer of the Turkey *Lonely Planet* guidebook), a Belgian lady in that one, an English woman in that, a Dutch couple in that one at the back, down the hill a bit is the American couple....", and so he went on (a decade later he would have been able to add my own house to the list of foreigner-owned houses in the mahalle): "This is now *Yabancı Mahallesi* (Foreigner Neighbourhood)!", he exclaimed. He then went onto explain that although the foreigner-buyers were partly to blame for rising house prices, he liked the fact that the old houses were being restored and lived in and he frequently helped the *yabancı* to organise the permits from the 'Protection Office' necessary to undertake the restoration work.

Figure 5.3 Fence around construction site in old neighbourhood.

During the 2000s, also, increasing numbers of cave-houses in the old ma-halles were converted into tourist accommodation. As discussed in Chapter 3, there was a fervour to turn houses into hotels at this time, largely due to the understanding that owning an accommodation business was the most legitimate, and effective, way to earn money from hot-air balloon commis-sions. Consequently, the old mahalles became dotted with construction sites, as houses and existing hotels underwent conversions, renovations, and expan-sions. While some house-to-hotel conversions occurred as the previous house-owners sold up and left the mahalle, others were developed by members of the owning household, which too necessitated the other household members moving out to live elsewhere. Consequently, while gentrification processes and displacement of residents from neighbourhoods are often understood as prompted by 'outsider' developers, be they private investors or government authorities, in Göreme's old mahalles, there have been several cases of moth-ers being displaced by their sons. At this time, the '*money-maker balloons*', as I called them in Chapter 3, began to have a profound effect both on family relationships and on Göreme's neighbourhoods more broadly.

Households that relocated from the old mahalles tended either to build a new house in the lower part of Göreme – usually Yeni (New) Mahallesi – where it was permitted to build anew, or they moved away from Göreme altogether. Hava's family, like many others from Aydınlı Mahallesi, sold their cave-house in 2010 and moved out of the mahalle. It was their next door neighbour who, having already converted his original family cave-house into

Figure 5.4 Hava's mother, Şerife, setting off to garden on donkey, 1997, in front of
        what is now Aydınlı Cave-Hotel.

a small 'boutique' hotel, bought Hava's family's house in order to conjoin the
two properties and thus create a larger hotel. A year later, he bought the house
to the other side to do the same. Right in the heart of Aydınlı Mahallesi, with
its 20 cave-guestrooms and a breakfast room situated on the top terrace with a
commanding view over the mahalle and beyond, this hotel, which first opened
in 2008, was aptly named 'Aydınlı Cave-Hotel'. The hotel's lobby was situated

Figure 5.5 Image taken in 1997 of Şerife (Hava's mother) in cave-stable, now the lobby of Aydınlı Cave-Hotel.

Figure 5.6 Aydınlı Cave-Hotel.

in the restored lower cave-stable where Hava's family had previously kept their donkey and cows. Meanwhile, Hava's family, who moved into the upper floor of a large modern house built by relatives who were recent return-migrants from Holland, were no longer able to keep any such animals. While their new house – situated adjacent to one of Göreme's central roads – was still close to Aydınlı Mahallesi, Hava and her mother were no longer 'neighbours', nor able to practice 'neighbouring', in the mahalle. In any case, Hava married and moved away to her husband's village, joining her new mahalle there as a *gelin* (incoming bride).

The loss of neighbours from the old mahalles due to people selling and leaving developed a momentum at this time; having fewer and fewer neighbours meant that others were soon keen to follow. The compounding effect of displacement has been discussed previously by Doucet (2009), who pointed out that, not only can 'the fear of being displaced...cast a spectre in the minds of [remaining] residents' (p. 301), but 'those who survive displacement pressures and are able to remain in their community may be at a loss because much of the community that built these networks will have been displaced' (p. 303). In my notes from a short fieldwork visit in 2013 I wrote about, on my first day, going around and visiting the few remaining households in the mahalle immediately around Hava's old house. The elderly women I visited were the mothers of the girls Hava was friends with when she was a girl, so I knew them from the times when I used to join their 'neighbouring' practices

Figure 5.7 View of Aydınlı Mahallesi in 1997 from the top level of Hava's house. The triple-arched building is the Aydınlı Camii/Mosque, dated 1883.

Figure 5.8 View of Aydınlı Mahallesi in 2019, with most houses converted to hotels. Photograph taken from the top terrace of Aydınlı Cave-Hotel.

during my 1990s fieldwork. My first neighbourly visit was to a house where 'Mother', as I called her, had earlier informed me she would cook my favourite dish – *kuru fasulye* – for lunch. To add to the *kuru fasulye* dish, Mother went down the outside steps at the back of the house to fetch some *turşu* (pickles) from the cave depot under the house. As we ate our lunch, we chatted about the changes that had occurred, including which households had left the mahalle since my last visit. At times during our chat, there was construction noise coming from the next door house; it too was in the process of being turned into a hotel. After lunch, we went out to the street behind Mother's house where a relative had earlier dropped off a large pile of pumpkins harvested from a family field. Mother resumed the work she had already begun that morning, opening, and de-seeding the pumpkins (the seeds she would later roast for snacking on during the winter). The elderly woman from across the street, one of Mother's few remaining neighbours, was sitting among her own pile of pumpkins doing the same. After a while of sitting with Mother helping her with the pumpkin work, I left and went around the corner to visit another elderly woman, who told me she had recently sold her house and would soon be leaving the neighbourhood. She had sold to the neighbour from across the road – the owner of Aydınlı Cave-Hotel – who had already bought and incorporated Hava's old house, plus the house next to that, into his hotel, and now wished to expand his business even further.

Some hoteliers' desire to expand their already-existing hotels led the old mahalles to become like the *Monopoly* board-game, whereby business owners

Figure 5.9 Boutique cave-hotel constructed from several old houses.

over time bought up clusters of neighbouring houses in the same street so that they could be linked to form a larger hotel. A prominent example in Aydınlı Mahallesi was Kelebek Hotel (previously Kelebek Pension) which, by 2008, had combined six neighbouring houses, and continued after that to add more. During the same period, in the other old mahalle – Gaferli Mahallesi – a larger hotel incorporating several houses was developed by a local businessman originally from that mahalle. Once the example had been set by these two hoteliers, others followed and several hotels in the old mahalles, including Aydınlı Cave-Hotel, were continually extended by knocking through walls and joining together what previously were separate houses. Remaining residents of these mahalles complained that such developments were ruining the mahalles irrevocably because, whereas a pansiyon or hotel converted from a single house could always be turned back into a house, a hotel developed from a larger cluster of houses could not – at least not so easily – be returned to residential use. Hence, the joining together of old houses in the mahalles for converting into hotels was seen as an act of irreversible change.

As well as being a trend-setter by incorporating more and more houses, Kelebek Hotel which was situated just up the hill from Hava's old house in Aydınlı Mahallesi, was considered a success prototype with regard to developing facilities and services in line with a more 'boutique'-style of hotel accommodation. In 2005, of the 50 or so cave-house accommodation establishments in the old mahalles, only 8 were fully converted into boutique hotels, while the majority remained as lower budget accommodation; at that time, over 20 establishments were aimed at low-budget backpacker travellers and still had dormitory rooms and shared bathrooms. By the end of the 2000s, however, these figures were reversed so that very few accommodation establishments aimed at backpackers remained. For hotels to have a restaurant terrace where evening dining and a breakfast buffet were served rapidly became the norm. Furthermore, following the owner of Kelebek Hotel's lead, other hoteliers in the old mahalles restored and refitted the cave and stone-arch rooms of the original houses so as to provide both a vernacular design aesthetic and a 'luxury' experience of marble fittings and jacuzzi bath in the ensuite bathrooms. Some hotels also added facilities such as a spa or a swimming pool, perhaps with a poolside-bar.

The hotels thus developed a sense of becoming 'tourism enclaves' in the midst of the mahalles. Although many of the tourists staying in these boutique hotels were the connection-seeking tourists, discussed in Chapter 4, who enjoyed wandering with their camera exploring the mahalle and viewing the daily life of its residents, unlike the backpackers who had previously stayed in the simple pansiyons in the mahalles, these tourists tended to spend a lot more time in their hotel; hanging around the poolside during the day, following this with pre-dinner drinks on the hotel's roof-top terrace watching the sunset, and then dining in the hotel's restaurant. Also, they tended to be transported directly to and from their hotel; for example, having transfers organised by their hotel from and back to the airport, and being picked up from the hotel when going ballooning or on a day-tour. Consequently, even though the hotels were positioned right in the middle of the mahalle, they became spaces that were thoroughly separate from mahalle life; even more so than the earlier backpacker pansiyons had done, they created a sense of being like islands. The sense of separation increased for Göremeli people as well as tourists. Previously, men were always popping in and out of each other's pansiyons, to have tea or a beer with their friend the owner. With the increased formality in structure of the boutique hotels – such as having a manager and designated housekeeping and kitchen staff – they had become places which the local men no longer recognised and where they themselves were not recognised.

A sense of dispossession, or 'place-based displacement' as Cocola-Gant (2018) calls it, became increasingly evident in the old mahalles towards the end of the 2000s. This was at least in part related to the processes of gentrification occurring, or, in other words, a change in class composition of the neighbourhood (Sakızlıoğlu, 2018). Gatherings for 'high society' parties and concerts became common, with the increasingly wealthy hoteliers hosting and mingling with outsider "*sosyete*" (high/polite society) Turks, and foreigner residents. Meanwhile, the remaining mahalle residents who had not (yet) succeeded in tourism business, a handful of whom might have been described as poor, watched-on as the mahalle changed around them. The dissonance between tourism and mahalle life appeared most striking one evening during a visit in 2010 when I was invited to attend a classical music concert to be held on the restaurant terrace of a boutique hotel in the upper part of Aydınlı Mahallesi. Sponsored by a local regional wine company, the concert would show-case regional wines alongside the "world-class" musicians set to play. It was to be attended by "*sosyete*" guests, both Turkish and international, including people of tourism importance, such as the then current and previous travel writers of the Turkey *Lonely Planet* guidebooks. It was a warm evening and on my way to the concert, I passed a group of mahalle women sitting outside on a doorstep. I chatted to them and told them I was on my way to the concert. They said that, since the concert was for "*sosyete*" people, it was not for them, but in any case they would be able to hear the music over the restaurant wall. Later in the evening, as I sat listening to the grand piano and string quartet, I wondered what the ladies sitting on the doorstep were making of the *sosyete* music playing in their neighbourhood.

While overt tensions directly between mahalle residents and tourists were unusual, the increased number of hotels directly neighbouring houses inevitably had implications for how mahalle life was experienced, and could be practised, by the remaining residents. Along with the increased presence of tourists in the mahalle streets, a daily influx of hotel workers, construction workers, service vehicles, and other tourism or construction-related comings and goings also increased in volume. With these comings and goings, the mahalle streets were inevitably rendered more 'public' or 'open' – as opposed to *kapalı* (enclosed) as they previously were; the mahalle was no longer the comfortable space of familiarity it used to be. Mahalle women were compelled to spend more time inside or at least within the walled courtyard of their house, as well as to wear more secure head cover when outside in the street. In chatting to an elderly woman whose house, nearby Hava's old house, had become completely surrounded by hotels, she complained to me not only about the increase in people and traffic in the mahalle streets, but also her sense, because of all the hotels surrounding her house, of feeling exposed and looked upon even within her home – such as when doing work on her balcony and roof terrace. She also said there were more water and electricity cuts in the remaining mahalle homes as hotels increasingly drew on the available resources, and she spoke of the now constant arguments in the nearby streets over parking places. In my 2009 fieldnotes, I recorded talking with some Australian tourists staying in a hotel in Aydınlı Mahallesi who told me they had experienced an incident earlier that day where an elderly woman in the street had thrown stones at them. It was clear that tensions and clashes in the mahalle were growing.

The mahalle as space of tourist consumption

Between 2011 and 2019, the mahalle streets became increasingly filled not only with ever more hotels but other types of tourism businesses also, particularly restaurants, tourist shops, and bars. The trend to open a restaurant business in the old mahalle began with Mehmet (introduced in Chapter 2) opening the large restaurant where the "*sosyete*" concert was held, and soon after by

Mustafa (also introduced in Chapter 2) opening in 2011 his cave-restaurant in the lower-middle (Orta) section of Aydınlı Mahallesi. By 2015, the number of hotels and restaurants in the network of narrow winding streets grew so large that there no longer was enough space at street corners to put signs up for them all, so the *Belediye* (Municipality Office) erected signs throughout the mahalle which had a number system to guide tourists to the particular hotel or restaurant they were seeking. I heard some tourists commenting that the signs made the place appear like a holiday village.

As well as the increase in the number and type of tourism-related businesses opening in the old mahalles, the growth of social media at this time also significantly affected the way in which the mahalles morphed into spaces of tourist consumption. As described in Chapter 4, the old mahalles became caught up in the contemporary 'selfie' visual culture, so that tourists would wander in the narrow winding streets with the sole purpose of seeking out aesthetically pleasing old stone houses (most of which were now converted into hotels) and their picturesque doorways as backdrops for their photoshoot. In this process, couples or small groups of tourists – often in photogenic 'costume' and sometimes with a professional photographer in tow – would be so thoroughly absorbed in their photoshoot activities that they had an apparent disregard for what was right in front of them, including the mahalle as a place of living for its inhabitants; they would 'see' only the aesthetic backdrops that might be used to create the perfect self(ie). As Smith (2018) remarks: 'Such performances on *Instagram*, innocent and certainly aesthetic as they appear, are indicative of an imagined possession of the land surveyed…, in almost all cases privileging the tourists' enjoyment over respect for longstanding local patterns of life' (2018, p. 181).

Hotels in the old mahalles became especially caught up in the 'selfie' visual culture and 'like economy' (Gerlitz and Helmond, 2013). In the earlier years of Göreme's tourism, the old houses in the top parts of the mahalles, many of which were long-abandoned, tended not to be converted into hotels since their position, being a substantial uphill walk from the restaurants and shops of Göreme's main street, was considered a disadvantage compared with the lower, more centrally positioned hotels. However, the effects of the *Instagram balloons* together with the 'like economy' prompted a repositioning of the most desirable hotels so that those situated in the top sections of the mahalles, such as Kelebek Hotel, became the most 'liked', and were fully booked throughout the year (see Figure 4.4). Illustrative of the changes in the desires of tourists is the way Kelebek Hotel's website described the hotel in 2019, two decades after the 1998 description quoted earlier:

Kelebek Hotel was built in 1993… It first started as a guest house and had 4 rooms available. Over the years Mr. Ali Yavuz expanded the guest house by combining it with the other houses around it…and currently, Kelebek has 35 rooms. Some of the rooms of the hotel are very unique as they were carved into the caves...Kelebek has stunning views of the valleys and the town...our guests can watch the hot air balloons from most of the rooms and terraces. Kelebek has a swimming pool, a Turkish bath & sauna, an a la carte restaurant...Kelebek offers concierge service where the hotel guest can book all the activities including balloon rides through the reception...Our sister hotel Sultan Cave Suites has a famous rooftop for balloon watch where Kelebek's guests have access to.

While the already elevated position of the hotels in the upper parts of the mahalles meant they could easily have a breakfast room with a view and a decorated roof terrace from which tourists could photograph themselves with the balloon-filled sky in the background, it was more difficult for hotels located in the lower parts of the mahalles to have these desirable features. Many of these lower hotels hence added one or two extra floors, usually in the form of a 'panoramic' breakfast room plus a roof terrace on the top. Consequently, there was not only a race for 'likes' in the 'like economy', but also a height race whereby more and more floors and roof platforms were sequentially added to neighbouring hotels. Moreover, as the new floors were often built out of wood and had panorama windows, they did not meld with the vernacular stone architecture and thus substantially altered the appearance of the old mahalles.

The ever increased height of hotel terraces and breakfast rooms also significantly affected the privacy of any remaining residents throughout the old mahalles; any courtyard could now be looked over and into, not only when balloons flew directly overhead in the early mornings but also when tourists

Figure 5.10 A hotel's added floors with breakfast room and roof terrace.

were eating breakfast, or doing *Instagram* photography, on nearby hotel roof terraces. Like Aydınlı Mahallesi, the old cave-house neighbourhood of Gaferli Mahallesi became similarly filled with boutique hotels and restaurants. As the ridge known as Sunset Viewpoint, situated just above Gaferli Mahallesi, became increasingly popular both for watching the sunset and for *Instagram* photography in the early mornings, the upper section of the mahalle came to be seen as a particularly good hotel and restaurant location. Householders increasingly sold up and left and, in 2019, there was only one residential household – an elderly couple – remaining in that upper part of the mahalle. The couple appeared defiant in the face of what was emerging all around them. For example, they continued to use a horse and cart for transport to and from the house despite the keeping of animals such as horses and cows no longer being permitted in the old mahalles due to their smells being deemed problematic for the numerous restaurants that had opened nearby. Meanwhile, the elderly couple's comings and goings with the horse and cart appeared to be of interest to tourists able to watch them from the terraces of surrounding hotels and restaurants.

As the number of hotels, restaurants, and shops in the old mahalles grew, not only was there an ever-increasing number of tourists walking up and down through these mahalles, but the number of workers – many of whom were from out of town – coming into the mahalles every day also grew. I was told in 2019 that approximately 700 workers came from nearby towns to Göreme every day to work in tourism businesses. One hotel worker from Nevşehir I spoke to as she was putting out the breakfast buffet told me that there were now several 7 a.m. buses from Nevşehir to Göreme, and that they were full – of mainly women – coming to work. Each day, all the workers would get off their bus and walk up through the mahalles to the particular hotel or other business where they worked. Between 2015 and 2019, also, a significant number of refugees, mainly from Afghanistan, were employed in the hotels and restaurants. They, too, would bus in from nearby towns each day, although some shared rental accommodation in Göreme's Yeni (New) Mahallesi, and they too would walk up through the old mahalles each morning to go to their place of work. A few Afghani men were given accommodation in the hotel where they worked, which meant they became neighbours, of sorts, within the mahalle. Late one evening, I was out in the street in front of Abbas and Senem's house helping them clear away after the day's *pekmez*-making work, and a young man came walking along the road. He did not greet us as he walked by and neither did we greet him. I asked Abbas who the man was, and he replied "Afghan". I asked how he knew that, and he responded that if the man had been Göremeli, he would have stopped and chatted; if he was Turkish but unknown to us, he would have said "*Iyi akşamlar*" (Good evening); as he did not say anything, he must be Afghan. He went onto say that, these days, out of ten people who pass by, only one will be from Göreme. This statement signalled a clear indication that the mahalle was no longer *kapalı* (enclosed) space.

While the workers in Göreme's tourism businesses were increasingly from outside of Göreme, business ownership – and especially that within the old mahalles – continued to be largely in Göremeli hands. In 2019, business owners I spoke to estimated that 90% of the hotels in the mahalles were owned by Göremeli people, with many of those being the original owner of the (first) house from which the hotel was built. Many of these hotels were no longer operated by the owner, however, since hotel owners increasingly decided to rent out their business. With many of those renting a hotel business being from out of town, this practice meant ever more 'unfamiliar' neighbours in the neighbourhood. For the property owner, though, with the going rate to rent a modest hotel in 2018–19 being 25,000 euros per annum, this made for easy income. Ownership of a hotel in Göreme thus became an easy key to wealth, as did ownership of any property in Göreme with prices having increased exponentially from 2000 onwards. One man who continued to live with his elderly mother in his family's cave-house in the middle of Aydınlı Mahallesi told me that, 20 years earlier when his family's cave-house was "worth nothing", he would never have imagined that one day his family's house would be worth a million U.S. dollars. That conversation took place in 2015 and, as the house was later sandwiched between two hotels, it likely became even more valuable in the years following. I was told in 2019 by the owner of Aydınlı Cave-Hotel that property prices in Göreme were rising so quickly that they were becoming comparable with property prices in Istanbul. As property values increased, every centimetre of a property was deemed valuable and worth (re)claiming and protecting. Tensions among hoteliers, and between hoteliers and the few remaining residents, ran high as they fought over property boundaries, including any cave-room space underneath which often ran beyond the above-ground boundary of the property it ostensibly belonged to. At times there were ferocious arguments between 'neighbours', creating an often-fraught atmosphere in the mahalles.

The inflated property values also created tensions *within* remaining households, as well as among families who owned the now crumbling cave-houses which they had abandoned after moving years earlier to an *afet* house in Yeni Mahallesi. Many of the households still resident in the old mahalles towards the end of the 2010s remained so only because of unresolved disputes among family members over their part-ownership of the old family house. Now that the properties had become worth so much, even those family members who had long since left Göreme having previously given away their part-ownership, returned to reclaim a stake in the property; some families were caught in a deadlocked feud over who owned what portion of the family property. Other families, among whom joint ownership was clearer, disputed whether to sell their house or to hold on to it and wait for its value to increase even further. Some remaining residents would have liked to turn their house into a tourism business, but were unable to do so because they lacked the cash needed to convert the property; despite their being wealthy on paper because

Figure 5.11 The narrow streets of the old neighbourhoods in 2019 had become busy
      places of tourist consumption.

of ownership or part-ownership of a valuable property, the few remaining residents in the mahalle tended otherwise to be poor in terms of their day-to-day
living. Furthermore, as with the three women who featured at the start of this
chapter, most of the remaining residents were elderly. They had seen much
change over the years, living in the midst of ongoing displacement and dispossession of their mahalle; no longer were the mahalles the domestic spaces of
familiarity and belonging they used to be.

The spectre of displacement

Having always loved the late afternoons in Göreme, when the day's work is
mostly done and it is time for *komşuluk* (neighbouring), I make a point of
walking around the mahalle at that time each day, to see who is sitting outside
on their doorstep or who else is out walking around. Although the chance of
encountering neighbours to chat with is becoming increasingly diminished
these days, I would still often find the now elderly Pembe, mother of another
of Hava's friends from those earlier years, sitting on her doorstep. As I perched
next to her on the step, we would keep our legs tucked in because of the cars
and minibuses rushing past up the narrow street. We needed to be particularly
mindful of the large tyres of tractors trundling by, on their way to deliver water
to the hotels in the upper part of the mahalle. When I made my first *komşuluk*

walk during my stay in the summer of 2019, I wandered up the street and sat on the step next to Pembe. After a welcoming embrace, she started to tell me how hard her life was becoming; she said her husband was sick and could barely walk now and, pointing to all the new restaurants up, down, and across the road from her house, she added that, because she only had three neighbours left, she felt alone, lonely (*yalnız*).

After sitting on the step for a while, I walked further up the street to visit 'Mother'. As was usually the case these days, she was sitting inside her house on her own when I arrived. She told me that because she only had one or two neighbours left now, she was thinking about moving to Avanos. There she would get a little house with a garden and live there more quietly, "and with neighbours". She said she would give her Aydınlı Mahallesi house to her sons to do what they liked with it; "They can turn it into a hotel, or whatever they want". Another afternoon, I bumped into the American couple who for many years had owned a house up the small lane opposite Pembe's house which they came to stay in for a couple of months each year. They, too, told me that they would soon sell and no longer come back to Göreme; soon their house would be "enclosed by hotels", surrounded on all sides, they said, and so for them "Göreme is wrecked". The conclusion that life in the mahalle was no longer what they had bought into had already been reached by most other foreigners who also had previously owned a house in the mahalle; most of them had already sold up and gone.

Life had become difficult for remaining residents in many ways. There was increased demand for water as hotels' swimming pools, flower gardens, and jacuzzi baths took a larger and larger share, meaning regular water-cuts in residents' homes and especially for those living in the upper parts of the mahalles. Until the late 2000s, residents had been able to access water at communal taps (*çeşme*) dotted through the mahalles but, by 2019, even those had dried up. The electricity supply was also over-loaded, and while hotels had generators which kicked in during power outages, residents had to cope with regular power-cuts. In addition, the early-morning rumblings as vehicles towing balloon trailers bounced up and down the cobbled mahalle streets, as well as minibuses picking up and dropping off tourists at hotels all hours of day and night, meant the noise and traffic associated with the hotels and balloons continuously filled the mahalle. There were also, in 2019, several new bars opening in the old mahalles; 'wine bars', 'cocktail bars', and 'roof bars', all offering "good music" and views of the sunset.

The filling of the mahalle streets with tourism-related activity led to an ever-increasing sense of dispossession, or 'loss of place' (Cocola-Gant, 2018, p. 288), for remaining residents who had lost 'the resources and references by which they define their everyday life' (ibid., p. 289). That sense of dispossession became startlingly clear to me on another of my afternoon visits to Pembe's doorstep. We were chatting as usual when suddenly the young woman from the house opposite came rushing out in a panic, saying her daughter was missing. She could not find the three-year-old anywhere and was clearly afraid for her safety. It was not long after that the girl's father arrived with his daughter sitting on his motorbike. He had come by a short while earlier and taken her to the shop down the road to buy sweets. The man was promptly chastised by his wife for his giving her such a scare, but the point that the incident did give her such a scare highlighted that the mahalle was no longer the space of belonging and collectivity it used to be; the comings and goings of strangers meant that the mahalle streets could no longer be considered *kapalı* space. With 'the fluid private/public home/street boundaries of traditional *mahalle* life' (Mills, 2007, p. 350) having previously been an integral aspect of daily life in the mahalle, the decreased sense of the mahalle streets being an extension of the interior space of the family is a potent example of tourism's deep mutation of place.

Many people told me that the Yeni (New) Mahallesi was now the only "real" mahalle in Göreme; "real" because it was where *komşuluk* (neighbouring) practices continued. The old mahalles had become like a "*tatil köyü*" (holiday village), or even a "*çizgi film*" (cartoon film), they said. In the *tatil köyü* mahalles, along with the actual departure, or displacement, of residents and hence the emptying of the neighbourhood of neighbours and neighbouring practices, there was a sense for any remaining residents that the mahalle was no longer theirs. Indeed, since the bonds of knowing (*tanımak*) were now lost from the mahalle, perhaps it could no longer be considered a mahalle at all: 'For…what does the mahalle become if it is no longer the space of the familiar?' (Mills, 2007, p. 351).

Notes

1 To whom I gave the pseudonym 'Esin' in *Living with Tourism (1)*.
2 I accessed this report in Ankara during my PhD fieldwork in the late 1990s.

6 Frontier of change

From being *kapalı* to crafting new selves

The frontier [of change] is a series of historically nonlinear leaps and skirmishes that come together to create their own intensification and proliferation. As these kinds of moves are repeated, they gain a cultural productiveness even in their quirky unpredictability.

(Anna Tsing, *Friction*, 2005, p. 33, [...]my insert)

Fieldnote diary 2019: Zekiye sells self-made jewellery from one of the wooden stalls situated along Canal Street in Göreme's centre: While I sat chatting with her at her stall one late-afternoon, we were joined by her daughter-in-law Emine. Emine had come from the Pharmacy down the street where she works; when she finishes work there each day she joins her mother-in-law at the stall so that, with her small amount of English, she can converse with passing tourists and help make a sale. Zekiye, who is in her mid-forties, told me that as well as speaking no English, she is barely literate; "They didn't want girls to go to school when I was young", she explained, "they wanted us to learn to make carpets. I went to school for three years but I didn't learn much". She continued: "I know most letters, but cannot put them together to read words". To demonstrate she attempted – in good humour – to read the writing on the side of a van that was passing, but she did not manage to read the words before the van was gone.

Born in the mid-1970s, Zekiye's three years of attending primary school would have been in the 1980s, just when tourism was beginning in Göreme. She went on to tell me about how her own life had been hard, relative to Emine's generation: "Women used to have a very hard life; they just worked in the house all day. They could not go anywhere, they needed *izin* (permission) to do anything"! She then talked about how her mother's generation had it even harder:

My mother had a *really* hard life. At the age of 15 she was married into a poor family in another mahalle. She had never set eyes on

DOI: 10.4324/9781003011200-6

her husband before the day of their marriage. Her mother-in-law was really nasty to her and didn't feed her properly. Sometimes she would escape and go back to her parents in her old mahalle to get some food.

Becoming increasingly reflective through our conversation about the lives of different generations of women in Göreme, Zekiye concluded that her life was relatively okay; "Now we earn our own money so our life is completely different from that of our mothers. The jewellery stalls give us özgürlük (freedom), and we earn our own money so our husbands cannot control us". Continuing, she said: "Anyway, the men have changed too. They have become more open-minded... it's completely different from before".

Figure 6.1 Women's craft stalls provided by the municipality (*belediye*) on Canal Street.

A remarkable aspect of my longitudinal study in Göreme has been observing the extent to which the frontier of change and life opportunities for Göreme women and girls has shifted. The middle-aged and older Göreme women I have met over the years of my conducting fieldwork had limited educational opportunities and have spent their life undertaking reproductive work in the domestic sphere. Even those who themselves were teenagers when I conducted fieldwork in the late 1990s had considerably fewer life choices

Figure 6.2 Zekiye and her mother, Hatice, at their craft stall.

than today's girls and young women. In *Living with Tourism (1)*, I depicted a gendered spatial and social duality in Göreme whereby, with their lives and practices spatially centred in and around the household and their immediate mahalle, women generally could only access the tourism economy indirectly through the earnings of their husband or other male family members. Twenty years later, whilst the older generations still revere notions of *kapalı* (introduced in Chapter 5 in relation to mahalle space, and meaning covered or enclosed) as the proper way for women to be – and the proper and *familiar* way for the mahalle to be, their granddaughters are able to go out and move around 'uncovered' (without a headscarf), and many these days leave Göreme to attend university. Göreme women also have become increasingly involved in tourism during the past two decades, gaining employment in hotels and guest houses during the 2000s, and increasing numbers embarking on entrepreneurial activities such as making and selling jewellery, and running shops or restaurants.

This chapter focuses in on this shifting frontier of change to look both at the role that tourism has played in the stretching out of spatial and moral boundaries in Göreme, and at how the 'series of historically nonlinear leaps and skirmishes', as Tsing (2005) puts it, have been negotiated in – especially women's – daily lives. Whilst gender relations in Cappadocia undoubtedly assert a definite women/men binary (see Tucker, 2003, 2007), focusing in on how notions of what is possible have shifted, especially for girls and women, allows for a view of gender roles and relations as fluid and negotiable, as always in a state of flux, and being reworked so as to reconstruct notions of what is appropriate or doable. In other words, whilst heeding Aitchison's (2005) suggestion to keep in sight the 'material constraints in women's everyday lives' (p. 219), a focus on how the frontier of change has shifted – whether out of necessity or out of desire – is to highlight 'the potential for the reworking, disruption, contestation, transgression and transformation of the dominant codes and behaviours of society' (ibid., p. 217; see, also, Tucker, 2022). Such reworkings and how they are negotiated are the focus of this chapter. Another vignette from my 2019 fieldnote diary further illustrates this frontier of change:

Fieldnote diary 2019: While walking up the street one afternoon, a loud thunder clap and the onset of heavy rain led me to duck into Emine's shop for cover. Emine was sitting chatting with another young woman and they pulled up a stool and invited me to sit down to wait out the storm. Emine introduced her friend Deniz to me and we quickly ascertained that, while we did not know each other, we had possibly 'met' during my fieldwork in the late 1990s – when Deniz was a baby.

I had not known Emine either from my earlier fieldwork, although I had become acquainted with her parents-in-law over the years and Emine and I met when I chanced upon her making *salça* (tomato puree) with her mother-in-law in the street and I stopped to help. Emine had grown up in the Antalya region on Turkey's south coast, but her family had originated from Göreme and they had returned in order for her to marry back into the community they had come from. Although she was living in an apartment in Nevşehir with her husband and their three year old daughter, as the '*gelin*' (incoming bride/daughter-in-law), Emine would come to Göreme every day, together with her husband and daughter, and spend her days based around the old cave-house of her parents-in-law. Her husband was working in a Chinese restaurant on the main street and, to supplement the household income, Emine had decided earlier in the year to open a small jewellery shop just around the corner from her parents-in-law's house.

Deniz, who is a year or two younger than Emine, grew up in Göreme and her mother is a long-time neighbour and good friend of Emine's mother-in-law; Emine and Deniz met and became friends after Emine married into the neighbourhood. In recent years, Deniz has been away at a university in the northwest of Turkey where she is studying business and accountancy. She was back in Göreme for the summer holidays when we met in Emine's shop, and when the new university year resumed, she would be in her final year of studies.

Not knowing most of the young Göreme women of Deniz's age, I asked her if many of her peer-group attended university. She said that these days, yes, many young women were going to university: "Because we want to do something with our life, do something for ourselves, rather than getting married early... For me, marriage can wait", she said. I asked her what she wanted to do when she finished her university study. She said that she would likely come back to Göreme and, initially at least, she would look after the accounts of her brother's horse-ranch business. Then she would perhaps work as an accountant for a hotel in Göreme and, eventually, she would like to start her own tourism business, perhaps a restaurant: "There are lots of opportunities in Göreme", she said.

During our conversation, the rain had stopped and after a while Deniz's mother, and soon afterwards, Emine's mother-in-law came into the shop to meet up with each other to go together to a neighbour's wedding.[1] Both of the mothers were dressed in their best clothes, although Emine's mother-in-law expressed embarrassment about her clothes being the more traditional *şalvar* (baggy trousers) and *yemeni* (village-style headscarf), while Deniz's mother's clothing was a more "modern" long skirt, town-style headscarf, and long overcoat. Both Emine and Deniz were wearing tight jeans and Deniz also a denim jacket; neither were wearing a headscarf. Also, neither of them were planning to go to the wedding. Instead, Deniz, at least, was going to visit a friend to wish them a happy birthday and would take a cake she had earlier gone by bus into Nevşehir to buy. I commented that 20 years ago, when I did my earlier fieldwork, I would never have known of a young woman from Göreme taking the bus into Nevşehir, dressed in jeans, to buy a birthday cake. The young women and their mothers chorused: "Of course – we were *kapalı* back then!"

Being *kapalı* and the first two decades of tourism

In *Living with Tourism (1)*, I explained how, in Göreme society, there was a clearly defined spatial separation of men and women in regard to both their economic and social activities (Tucker, 2003). This gender separation was in accordance with Islamic codes and practice and upheld by principles of shame and honour.[2] In reference to Turkish village society, Paul Stirling (1965) had earlier noted that prestige was very much 'dependent on honour, *namus*, and this was directly related to the women of lineage households' (p. 168). Stirling continued:

> An honourable man is …able to keep his women pure from all taint of gossip… The opposite of *namuslu*, honourable, are *namussuz*, without honour, or *ayip*, shameful. These two words are in constant use, mainly for reproving children or for critical gossip.
>
> (ibid., p. 231)

In line with this, I learnt during my 1990s fieldwork in Göreme that keeping women 'pure from all taint of gossip', in Stirling's words, was closely linked with the word '*kapalı*'. Being *kapalı* (covered or enclosed) denotes the proper, or honourable, way for a woman to be and refers to the covering of the hair with a headscarf, covering of the body including legs and arms with clothing loose enough to not show the contours of the body, and staying 'enclosed' within the household and immediate neighbourhood. It was deemed shameful, and would bring dishonour on the family, for a woman to be uncovered, and likewise for her to *gezmek*, meaning to travel, walk, or wander around, beyond the neighbourhood. If a woman wished to go out of the immediate neighbourhood – for example, to go to the market or to visit a friend or relative who lived in another neighbourhood – she would first obtain permission (*izin*) from her husband or father. In *Living with Tourism (1)*, I quoted from an interview conducted in 1997 on the meaning of the concept of *gezmek* for women in Göreme:

> They [women] can only comfortably *gezmek* in their street, they can't *gezmek* in the centre. Maybe if they want to get somewhere, they can

pass through the centre, but they cannot sit and drink tea, or sit and eat – they can't do anything. If they go out, it is always to a house – to their neighbours' house – to their friend's – always like that. They can *gezmek* comfortably like that only, nothing else.

(Tucker, 2003, p. 77)

While there had long been a primary school in Göreme, being *kapalı* meant that girls and young women tended to have less access to education than their male peers; as illustrated by Zekiye in the vignette at the start of this chapter who – being of primary school age in the 1980s – went to school for just three years. Although primary school attendance grew during the 1990s, most Göreme girls did not attend high school as that would necessitate travelling to Nevşehir and it was deemed by their families to be inappropriate for them to go out of Göreme. Although the 1980s and 1990s had seen a marked improvement in education at the higher level for girls in other parts of the country, the predominance of village life and conservative values prevailed especially in the central and eastern regions; around the turn of the millenium, the illiteracy rate for women in these regions, which included Cappadocia, was still around the 50% mark (Durakbaşa and Karapehlivan, 2018). The women and teenage girls I befriended during the 1990s had either left school at the end of primary school or, at best, middle school. The elderly women today – and most of the middle-aged women – are illiterate. The three elderly women I referred to at the start of the last chapter, whom I joined as they sat out on the steps of a small side street in the lower part of the mahalle, told me, like Zekiye above, how hard their earlier life had been. Two of them had had no schooling and told me that since they were illiterate they were only able to buy groceries at the bazaar/market because there they could talk to the stall-holder and ask for what they wanted; in the new supermarkets in town, they needed to be able to read in order to know what they were buying so they could not use those stores. I asked what they used to do when young if they did not go to school. They said they spent their days weaving carpets, working in the fields, and doing housework, such as cleaning, bringing water, and preparing food: "It was very hard", they chorused.

The point that being *kapalı* was normative behaviour for women in Göreme, as was maintaining a clear gender separation, was crucial also to understanding how gender relations would interact with tourism development in Göreme over the years. As tourism started to develop in the region, women were largely separated from tourists and tourism activity so that tourism business became almost entirely the domain of men. Even though significant change regarding gender roles and relations had already started to occur elsewhere in Turkey, Göreme and the wider Cappadocia region remained, during the 1980s and 1990s, a pocket of Islamic social conservatism (Tucker, 2003). Women's domain – social, economic, and religious – was firmly centred around the house-hold and immediate neighbourhood (*mahalle*), while men's domain was the *merkez* or *çarşı*; the central area where the municipality office, teahouse, market places and grocery shops, and the central mosque were situated. During the first two decades of Göreme's tourism, apart from a few pansiyons and cave-hotels dotted around the old mahalles, the majority of tourism businesses, including restaurants, cafes, souvenir shops, and tour agencies, were situated in the central area, strung along either side of the main streets. While women could pass through the central streets to get to the weekly market, or to go to the health clinic or pharmacy for example, they did not feel comfortable there, and would pass through the area quickly and without lingering.

In other words, Göreme women had no 'place' in the *çarşı* and, relatedly, in general, they had no 'place' in the tourism economy.[3] Almost all of Göreme's tourism businesses during the 1980s and 1990s were run and worked in by men, while women continued to engage in garden-agriculture and other household and reproductive duties. The mother of Deniz and mother-in-law of Emine in the vignette above have lived being *kapalı* and have done household and reproductive work all of their life; they have had little or no access to the paid economy. Like other Göreme women of their age, they seldom, if at all, travelled outside of the Cappadocia region. Their days would be spent doing garden and food production work, raising their children and, as their main form of permitted 'enjoyment', in the summer months, attending village weddings.

It is interesting, then, to consider what contingencies have coincided to enable the lives of those mothers' daughters to be so different; how has the necessity for women to be *kapalı* lessened so significantly in only one generation? Furthermore, what has the role of Göreme's tourism been, in relation to the broader context, in accelerating particular 'non-linear leaps and skirmishes'? Deniz, in her final year of studying business at a university in the north-west of Turkey, exuded a sense of independence and life opportunity when describing her student life and telling me of her future plans. Meanwhile, Emine chose to open a shop which, as well as supplementing the family income, gives her a place in Göreme's tourism economy *and* a place on the now-busy thoroughfare of *Aydınlı Mahallesi*. Zekiye, in the vignette at the start of this chapter, said that her jewellery stall on Canal Street afforded her a sense of "freedom". During the past two decades, a renegotiation and shifting of spatial and moral boundaries has enabled women not only to engage in higher and tertiary education, but also to negotiate a place for themselves in tourism business.

'Space invaders': women's 'place' in tourism work

In Chapter 2, I introduced Hanife and Hayriye, the two sisters who, during the 2000s, had first run a stall in a valley outside of the village and later that decade, opened a shop selling handicrafts on Göreme's main street. In a conversation I had with the two women in 2010, they told me how shy they had felt when they first opened the shop. They also told me, proudly, that it was due to "their bravery" in opening the shop that other women felt able to come to the *çarşı* (central streets); "It doesn't cause gossip anymore", they told me. Prior to then, the gendered spatial and moral boundaries had meant that women had no place in the *çarşı*; the fear of the 'taint of gossip' meant that they were not confident or comfortable being in Göreme's central streets, let alone working or doing business there. Indeed, as tourism initially developed, the restaurants, shops, and tour agencies located in the central streets rendered the *çarşı* even more 'public', or open, than it had been previously; not only were tourists present there but also men from outside of Göreme who had come for work or to do business in tourism. These points are summarised by women I interviewed in 2005, one of whom said: "I can't even go to the shops or the market, because of tourism": "My husband won't allow me to go to the *çarşı*, …Also I don't have the confidence", another woman told me. Hotels and guest houses – including those situated in the residential mahalles (neighbourhoods) – were, like the *çarşı*, generally considered inappropriate spaces for Göreme women to be.

However, the 2000s saw a marked increase in women's presence in tourism spaces and involvement in the tourism economy. Initially, during the earlier part of that decade, women generally worked in the 'backstage' sections of hotels only, doing cleaning and laundry work. With many hotels becoming larger, more formal and more compartmentalised "boutique hotels", the creation of a backstage space allowed more women to be employed. As one woman working as a cleaner in a hotel in the mid-2000s told me: "Here it is safe, we feel comfortable working here". A small number of younger Göreme women started to work in front-line jobs doing hotel reception work or serving in restaurants in the *çarşı* also, although this was prone to gossip and chastisement. During a chat with a group of women who worked as cleaners in a hotel, I was told: "Some restaurants have girls working and wearing short skirts – this is wrong". By the end of the 2000s, many more women were engaging in tourism entrepreneurial activity also, and while this was mostly still in the mahalles where women sold handicrafts to tourists they invited into their cave-house, a small number of "brave" women – such as Hanife and Hayrire – were starting to do tourism business in the central *çarşı* streets. Compared with my 1990s fieldwork, during fieldwork visits in the late-2000s, it was highly noticeable that many more women were present in the central streets of Göreme, often at their place of work.

In a sense, these women could be considered 'space invaders', a term used by Puwar (2004) in reference to women and racialised minorities beginning to enter spaces from which they were historically and conceptually excluded. Drawing on Massey's (1994) connecting of bodies and space, Puwar notes that some bodies have the right to belong to particular spaces while others are designated as trespassers, or as being 'out of place'. It is therefore an illuminating point of focus when 'trespassing' occurs, or when out of place bodies invade a certain space from which they had long been excluded. Puwar argues, in line with Tsing's (2005) frontier metaphor, that this focus on 'space invaders' is 'intriguing because it is a moment of change. It disturbs the status quo, while at the same time bearing the weight of the sedimented past' (Puwar, 2004, p. 1). A similar spatial reference is in Cone's (1995) depiction of female entrepreneurs in Chiapas, Mexico, who, in their craft production and sales to tourists, 'craft new selves' by 'stepping outside' 'onto public squares and thoroughfares, social settings that had previously been closed to them' (Cone, 1995, p. 315). Indeed, having 'stepped outside' to do tourism business in the central *çarşı* streets of Göreme, Hanife and Hayriye would similarly fit with Cone's notion of 'crafting new selves', as well as with Puwar's notion of 'space invaders'. As Hanife proudly stated, "We were the first ones to do business in the centre – we started it". "Now there are lots of women in the centre", she continued; "Now we can sit on the pavement at the front of the shop. Women *gezmek* (walk around) in the evenings, having ice-cream. We're getting used to it now".

Since Hanife and her sister made that first leap, the ability for women to access and participate in the tourism economy, including in the *çarşı* streets, continued to grow: "We opened it up for the others", Hanife told me. During the 2010s, more Göremeli women became entrepreneurs in their own right, including one woman who accessed a national 'women-in-business start-up' fund to open a restaurant, and Fatma – introduced in Chapter 4 for her 'connecting' with tourists – who started a cooking class business for tourists as well as managing the family restaurant on the main street. In addition, the wooden stalls provided by the *Belediye* (town council), where Zekiye in the vignette above displayed and sold her handiwork, were set up to allow Göreme women to engage in tourism business for themselves. While these women would often complain at the lack of sales they made, they acknowledged that having the council stalls provided rent-free allowed them to make a little bit of money of their own, plus it afforded them a legitimacy to be in the centre of town where they could socialise with each other and with passers-by. In other words, the stalls had allowed the women to 'step outside' onto public thoroughfares and hence to 'craft new selves'.

Zekiye said, also, that the 'opening' of women's lives had occurred more extensively in Göreme than in other villages and towns in the region because of tourism: "Here, tourism opened our eyes" she said, "including the eyes of Göreme men, because we saw foreigners and gradually we said 'We too want to be free like that'". The "foreigners" Zekiye was referring to, those who in a variety of ways introduced Göreme women to 'other' ways of being a woman, were in large part the many 'tourist' women married to Göremeli men; reportedly, by 2005, there were 38 such 'mixed-marriages' and this number continued to grow in subsequent years. These mixed-marriage families often deviated from usual Göreme *kapalı* propriety, with the 'tourist-*gelin*' (brides) tending to work in tourism, usually in the business of their partner or husband. Foreign *gelin*s working in tourism in Göreme was generally condoned because their dealings with and understanding of tourists' wants and needs were considered good for business, and so not only were they able to contribute to the household economy of the family they became a member of, but they often contributed innovative ideas which were picked up by other tourism businesses in Göreme. Overall, by demonstrating 'other' possibilities regarding gender roles and relations, the presence of foreign women working in Göreme's tourism businesses worked to loosen the boundaries of honour and shame imperatives and thereby shift gender norms.[4]

The reworking of the boundaries of honour and shame imperatives through women's 'trespassings' and 'space invasions' often occurred out of necessity rather than through a desire to craft new selves per se. For many Göreme women, the ability to contribute to the household income was significant. In other words, some households could only make ends meet if the women cleaned in a hotel, or worked in the family business thereby negating the need to employ a non-family member. Hence, economic necessity cut against the moral imperative for household women to be *kapalı*. An example was Fatma who ran cooking classes (see Chapter 4) while also cooking for and running the family's café on the main street. At times, Fatma appeared to find all of her responsibilities very hard and would much rather have not had to work in these various roles. In her case, the economic necessity propelling her to 'craft a new self' seemed to cut against desire altogether. Other women too, such as those who sold jewellery from their doorstep in the old mahalles, were often doing so to get by and because they felt they had little other choice; also, they often had to contend with gossip and criticism from their neighbours. However, these women's ability to earn their own money was an integral aspect of their sense of 'crafting new selves'; in my conversations over the past two years with women who worked or engaged in tourism business, almost all of them reflected on the importance for them of earning their own money. As Zekiye said in the earlier vignette, because she and the other women who ran handicraft stalls were able to earn some money, "our husbands cannot control us". Indeed, many women suggested that working in tourism had afforded them an increased independence and a heightened sense of self-confidence and self-worth: "Of course I have more independence because I don't have to ask for money from my husband. Therefore, as well as earning money, working in tourism affects our head and ideas". Hence, even if a woman's paid employment in a hotel or entrepreneurial endeavours had begun out of economic necessity, a sense of 'crafting new selves' tended to eventuate nonetheless.

The ability to stretch boundaries and craft new selves undoubtedly varied for different generations of women. Whilst some women I spoke to referred to their husband's role in allowing or disallowing them to work in tourism, many of the 'skirmishes' taking place in relation to this frontier of change occurred between generations, for example, through young women's grandmothers' attempting to keep their granddaughters *kapalı*. In speaking to an elderly woman, I was told:

When the young ones work in tourism, they lose their morality. A neighbour's daughter works in a hotel's reception and I wouldn't want that at all. She sometimes has to work until late. Women who mix with men a lot lose all sense of shame. They are not shy, not reserved, there's nothing left. This is what tourism does to people and morals.

The fear of gossip played a key part in negotiations regarding *kapalı* imperatives, and because this fear was felt most strongly by the older generations, younger families who did not live with their elderly (grand)parents were freer from the grip of gossip. A younger woman working in a hotel told me:

> The older people think that tourism is something outside of their religion. So there is still some shyness, especially in families with older people. They still follow the old culture. The families who don't have any older people are more comfortable with modern life. They don't care about gossip.

Nuclear family living arrangements became more prevalent throughout Turkey during the past two decades, and in Göreme, this trend was accelerated due to the tourism economy allowing newly married couples to rent or buy their own house, thus deviating from the traditional extended family arrangement in the paternal home. The woman quoted above went on to explain: "Families are getting smaller. Young people are moving out of the family house to live in their own house. This takes the social pressure by the old people off the young people".

While nuclear family arrangements allowed younger people to make 'leaps' of change with potentially fewer skirmishes, there is no doubt that women's entrepreneurial activities, along with the increased employment of women in hotels and other tourism businesses, still necessitated substantial spatial negotiations. In relation to her discussion of 'space invaders', Puwar (2004) refers to the importance of knowing how to behave, or even how just to 'be', in a particular setting. Puwar refers to this knowledge as the 'tacit requirements' – or 'soft things' – which enable engagement with the 'normative ways of being' (Puwar, 2004, p. 116) in a particular space or setting. Many of the Göreme women I interviewed referred to relative confidence and feelings of comfort or discomfort regarding their tourism activities and in relation to different settings: "My husband won't allow me to enter public places…Also I don't have the confidence"; "Now women are becoming more confident"; "Here it is safe, we feel comfortable working here".

In a tourism setting, the soft things, or tacit requirements, also likely include competencies more broadly associated with the notion of cosmopolitanism (Notar, 2008; Swain, 2009) or, as Hiebert (2002) describes it, a 'capacity to interact across cultural lines' (p. 212). Cone's (1995) earlier suggestion that, '"Stepping outside" requires that women create new *kinds* of relationships and new identities' (p. 315) similarly evokes the concept of cosmopolitanism. Over the years of their 'stepping outside', Hanife and Hayriye have had to develop what Vertovec and Cohen (2002) refer to as a cosmopolitan 'skill of manoeuvring through systems of meaning' (p. 13) in their capacity to act as interlocutors between women craft producers from other villages and international buyers for their products. Fatma, too, has come to understand that when leading a cookery class in her cave-home, she should negotiate representing a 'traditional' identity for tourists, and yet at the same time emerge from that identity enough to host the tourists comfortably in her home. Indeed, many of the women who invite tourists into their cave-home succeeded in being able to connect with tourists, some also forming arrangements with guides to bring their tour groups. Since tourism activity had entered what was otherwise considered domestic space, these spaces were reconfigured through tourism and so the gender identities they produced were also reconfigured.

As I have written in earlier chapters and elsewhere (Tucker, 2009a, 2011), however, the smooth-running of these interactions is somewhat precarious and, while sales might be made, the process of the sale can often result in discomfort both for the Göreme women and the tourists involved. Relatively isolated in the confinement of their entrepreneurial activities to 'domestic' space, the women have had limited ability to acquire knowledge of tourist imagination and expectation – in other words, to learn the 'soft things' (Puwar, 2004, p. 110) – that would enable them to successfully act in their interactions with tourists. As Swain (2009) has suggested, these embodied 'cosmopolitan' skills and knowledges can only be developed through participation in repeated and varied interactions with tourists. Whilst Göreme men – through their repeated interactions with tourists as well as by observing each other's successful and not so successful interactions with tourists – have developed a 'capacity to interact across cultural lines' (Hiebert, 2002, p. 212), Göreme women have only recently negotiated a 'place' for themselves in tourism; hence, they have had relatively limited opportunity to craft cosmopolitan selves.

Indeed, for many of the women who have attempted to negotiate a 'place' for themselves in tourism work or tourism business more generally, their endeavours have been a challenge. Fatma, for example, is curtailed in her ability to run the cooking classes on her own because of her limited English language ability, and likewise, when she is running the family café, she has limited ability to do table service. The women who have jewellery stalls on the main street also struggle in their communications with tourists and, consequently, they complain of having low sales. Likewise, Emine, who has a small jewellery shop in the mahalle, feels challenged in her interactions and communications with tourists. As Salazar (2010) has noted, the ability to become a cosmopolitan is inevitably structured by considerable inequality.

An initiative which, during the last two decades, did help in enabling Göreme's women to 'step outside' and negotiate a 'place' for themselves in the *çarşı* was the setting up in the early 2000s of education programmes for women in the *Belediye* (town council). The Göreme Mayor at that time had lived for many years in Ankara and, when he returned to his home place of Göreme to stand in the mayoral elections, his wife came too and brought her own ideas of how to improve the lives of Göreme's women. Once her husband was in the mayoral office, she secured funding related to United Nations 'Local Agenda 21' sustainability initiatives and organised women-only classes in literacy, numeracy, and English. That these classes were held in the *Belediye* building in the centre of Göreme gave the women who attended the classes a legitimacy to 'be' in the *çarşı*. In addition, the skills learnt in the classes gave women increased ability, as well as more confidence, to get a job in a hotel or other tourism business. The classes for women were also indicative of social reform occurring at the broader level which, in conjunction with the shifting of spatial and moral boundaries occurring in relation to tourism in Göreme, opened up other 'spaces of desire' in the broader context.

'Spaces of desire' in the broader context

Fieldnote diary, 2019: Hatice was born in the late 1990s, during the time of my earlier fieldwork, and when I met her in 2019 she was a 22 year old practicing clinical psychologist who also helped out a bit at the family's hotels by baking cakes and other local delicacies for the breakfast buffets. The day I met with her she didn't have any appointments at the clinic where she worked in Nevşehir, so she was spending the morning helping at the newly opened hotel of her father, Osman (introduced in Chapter 2).

Hatice told me that she had recently graduated after five years of psychology study at university in Kayseri, the regional city about an hour's drive from Göreme. While studying, she had spent two months doing an internship in Belgium, during which she visited Paris and Amsterdam. She had also spent two months staying with a family friend in Canada, with the main purpose of improving her English. She told me that when she was planning to go to Canada, her grandmother had argued with the family friend, saying "You cannot take my grandchild, I will not allow it"! Hatice explained that her grandmother was of a different generation: "She was a traditional woman and she didn't want me to go out in the evening. After the evening prayer, I couldn't go out of the house. It was very hard". She continued to explain:

But now things have changed a lot, I go everywhere, all of the time… When I went to university everything changed, because before that I wasn't free…and of course when you go away to university you will be out of this, you will be out of that control.

She continued, "And that was a good feeling for me. Because I was free. I didn't have to ask permission from anybody; it was good, because I was free".

I asked about her relationship with her father and his allowing her "to be free". She said:

> He wants me to be a free woman. He always supports me, all of my decisions, all of my choices. When my grandmother said something, my father always supported me. Yes, he was always between me and my grandmother. It was very hard for him. He was always in the middle.

She went on to tell me that her father had become open-minded because of tourism, and because of other families:

> We saw other families on television, or other families here such as our neighbours or my dad's friends, sending their daughters to university, and he saw that it was okay, it can be possible. So he would talk with my mother about me and say "Okay, let's try that this year... and so things changed.

I asked about other Göreme women of her age group and whether their situations were similar to hers. She said that, while some of her friends were similar, others were in different situations because they had not been to university: "They stay at home and work for their family, because they have to. They didn't go out of Turkey, they're not free".

When I asked Hatice about how she sees her future, she told me that she was making plans to obtain a master's degree, maybe abroad: "I'm working on that. My brother's helping because he always supports me to go 'up'. And my father does that too. I'm very lucky".

Young women such as Hatice, and Deniz from the earlier vignette above, are the first generation of Göreme women to attend university and go on to practice a profession. Over the past two decades, various local, national, and global 'fluxes', as Jimenez-Esquinas (2017) has called them, or what Tsing (2005) has referred to as 'historically nonlinear leaps and skirmishes', have 'come together to create their own intensification and proliferation' (ibid., p. 33). Skirmishes occurring during the 2000s and 2010s might have included fierce debates in Turkey's national-level upper echelons when European Union accession criteria clashed with conservative Islamic values. Or family-level skirmishes – equally fierce in intensity – such as when a Göremeli grandmother argued with a young woman's father over the 'leap' of allowing her granddaughter to go away to university or to spend time abroad. However intense and whatever the level, such 'leaps' – like women's 'leaps' to transgress boundaries and engage in tourism business – must firstly, somehow, arise as either *necessary* or *desirable*. On this, Tsing (2005, p. 32) adds to her discussion on the notion of 'frontier' by saying: "It is a space of desire: it calls; it appears to create its own demands; once glimpsed, one cannot but explore and exploit it". As Hatice suggested, seeing "families on television" – perhaps in Istanbul-based soap operas for example – gave a glimpse of what "can be possible", thus opening up a space of desire. Simultaneously, seeing neighbours and family friends sending their daughters to university enabled her own family to see that "it was okay".

Also opening up this broader space of desire, opening up what "can be possible", were significant social reforms at the national level related to gender equality ideals (Durakbaşa and Karapehlivan, 2018). Although Islamist pushes have periodically been re-exerted, including in recent years, such reforms have been ongoing since the early days of the secular Turkish Republic 100 years ago, so that generally, over this time, women in many regions of Turkey have experienced increased levels of equality and participation in the workforce (Kandiyoti, 2002; Müftüler-Baç, 2012; Sönmez, 2001). Considerable improvements in access to education for girls and young women have been made also (Durakbaşa and Karapehlivan, 2018; Müftüler-Baç, 2012); for example, the introduction in 1997 of eight years of compulsory education was important for raising girls' schooling rates, and it was at this time – the time of my 1990s fieldwork for *Living with Tourism (1)* – that Göreme gained a middle school alongside the primary school. Turkey's adoption of EU accession criteria in the early 2000s prompted further gender equality reforms, as did engagement with other international agendas such as the ongoing Local Agenda 21 initiatives. The introduction in 2012 of twelve years of compulsory schooling, including high school attendance for four years, led to the schooling gender gap becoming closer to closed in Turkey. In addition, Turkey achieved high levels of female students graduating from university during the 2010s, with "highly favourable percentages of women among the highly educated professionals which hint at gender equality" (Durakbaşa and Karapehlivan, 2018, p. 75).

However, as already mentioned, conservative values – including imperatives for women to be *kapalı* – meant that some regions continued to lag behind in terms of gender equality, including in relation to girls accessing education; as recently as 2016–17 in Central Anatolia – where Göreme is situated – only 66.8% of girls attended secondary school (Durakbaşa and Karapehlivan, 2018). A sharp increase in girls and young women from Göreme accessing secondary and tertiary education in recent years was therefore something of an anomaly in the regional context. An even more notable anomaly was that, during the 2010s, the rate of secondary and tertiary education among Göreme girls and young women reportedly overtook that of boys and young men, with the latter being more inclined to drop out of school earlier than girls. This localised anomaly is likely, at least in part, to be linked to particular tourism-related anomalies at the local level, whereby tourism-related fluxes and 'leaps' have opened up other spaces of possibility.

Perhaps the most significant tourism-related flux, or leap of change, in Göreme during the past two decades was increased wealth. Tourism earnings were undoubtedly an important factor in enabling Göreme families to meet the financial costs incurred for their children to access quality education, including private high schools, colleges, and universities. The Turkish education system was subject to extreme privatisation policies during the 2010s particularly (Durakbaşa and Karapehlıvan, 2018), and a large number of private schools and other educational institutions were established across the country, including in the Cappadocia region. While national scholarship schemes were available, many Göreme families became able – due to their increased wealth from tourism – to aspire to sending their daughters as well as their sons to private colleges and universities without having to rely on scholarships. They could also more easily afford the extra living costs associated with their sons and daughters living away from Göreme while they attended university.

As well as tourism enabling Göreme families to meet the financial costs of education, tourism encouraged an influx of more 'open' (as in, not *kapalı*) incomers who demonstrated 'other' possibilities regarding gender relations and roles; again, in Anna Tsing's (2005) words, creating 'a space of desire' which 'once glimpsed, one cannot but explore' (p. 32). Zekiye, with the stall on Canal Street, comes to mind once again in her saying: "Here, tourism opened our eyes, including the eyes of Göreme men, because we saw foreigners and gradually, we said, 'We too want to be free like that'". The many mixed-marriage families, referred to above in relation to the foreign *gelin* (brides) working in tourism, involved women from countries, such as the United Kingdom, Australia, New Zealand, South Africa, and Japan, who had originally come as tourists and later married into the Göreme community. These women were generally keen for their children – boys and girls – to have as extensive an education as possible, and the mixed-marriage families were among the first to send their daughters to private high schools and colleges in Nevşehir, as well as sending them away to study at university, either elsewhere in Turkey or in their mother's country of origin. Again, these families demonstrated, to use Hatice's words from the vignette above, what "can be possible".

The marriages between Göremeli and foreigners/tourists meant, too, that otherwise conservative families gained in-laws from and in other countries around the world; this in itself – although to varying extents – opened minds to other societies' gender norms. In addition to these foreign in-law relationships, some long-term friendships formed between Göreme people and repeat-visiting foreigners/tourist (or researchers such as myself), thus creating strong links with people in other countries. As with the example of Osman's daughter Hatice going to stay with a family friend in Canada, in some instances, these overseas friends became mentors, sponsors, and hosts to ease the way for Göremeli daughters and sons to visit and perhaps even attend secondary or tertiary education in a foreign country. That some of the more successful hoteliers and other businessmen in Göreme have followed these 'social capital' opportunities has resulted in setting new precedents; as well as glimpsing what else can be possible, copying or competing-to-remain-equal tendencies (discussed in Chapter 2) meant that others wished to emulate the 'leaps' of their apparently successful peers: "If he's sending his daughter away to university then so can/will I".

Along with foreign *gelin* (brides), 'incomers' included returning migrant families. These were Göremeli families who had taken advantage of migration programmes in the 1960s and 1970s which saw many people from Göreme migrating to work in Germany, Holland, and Belgium as well as Turkey's western and southern regions. Most had returned each year to Göreme during the summer months, but some families – drawn by the opportunity to invest in tourism business – decided to return for good. Several young women from such families married Göreme young men who themselves had become prosperous through tourism. During the 1990s and early 2000s – when high school and university for Göreme girls were still more or less unthinkable – these returning 'semi-Göreme' families, whose daughters as well as their sons were likely to have attended university, demonstrated that "it was okay" to extend education opportunities for girls and young women.

Referring again to the frontier of change metaphor, Tsing's (2005) suggestion that 'as these kinds of moves are repeated, they gain a cultural productiveness even in their quirky unpredictability' (p. 33) seems pertinent in the fact that, by the end of the 2010s, the rate of secondary and tertiary education among Göreme girls had reportedly overtaken that of boys. This is likely connected to boys and young men in Göreme coming to view tourism as a good, and perhaps easy, way to make a living. It is believed generally in Göreme that success in tourism business does not require any particular education or qualification; if a young man has grown up hanging around his father's or another relative's tourism business, he is likely to have picked up some English as well as to have gained a reasonable understanding of tourists' needs and wants. Hence, being keen to, sooner or later, work full-time in tourism – either entering their father's tourism business or generating their own business, young men developed a propensity to leave school earlier than girls. Simultaneously, for girls and young women, it was increasingly seen to "be possible" to get a higher education and even to develop a professional career. During my fieldwork in the 2010s, I met several Göremeli girls and young women who aspired to become nurses, accountants, doctors, and lawyers; as I will discuss in the next chapter, they had hopes of a better life, of 'becoming other'. Important in this was that, whilst boys and young men had long been able to see tourism as a viable option to make a living, girls and young women had not; the gendered spatial divide in Göreme meant that women, on the whole, had never really had a 'place' in Göreme's tourism, or at least, their 'place' there was limited. Hence, while for a few Göreme women tourism work or entrepreneurship had become a space of desire, or perhaps more often a space of necessity, a new space of desire which opened up for young Göreme women growing up amongst Göreme's tourism during the past two decades was further education and a career elsewhere.

Quirky unpredictability at the frontier of change

Women's initial 'space invasions' as they stepped out into the *çarşı* were brave leaps of change which, when repeated, not only incurred a 'crafting of new selves', but also gained a cultural productiveness which stretched spatial and moral boundaries outwards in the ever-shifting frontier of change. Meanwhile, the 'spaces of desire', and the ability to 'explore' them, expanded – for younger Göreme women especially – to include opportunities to craft new selves, to become other, elsewhere. One more vignette illustrates how these remarkable reworkings of 'what is possible' manifested in the daily lives of three generations of women in one household:

Since Neriman was widowed when her children were young, she not only needed to keep the household fed and clothed, but she also had to be the maintainer of *namus* (honour), both roles usually held by the man of the household. Neriman had come to Göreme as a *gelin* (bride) from a nearby village and, born in 1947 into a conservative family, she had no schooling as a girl and had never learnt to read. In 2018, she was encouraged by her grown-up children to attend literacy classes for women put on by the Göreme *Belediye* (Municipality office). She found the lessons hard and did not go for long, but she was pleased to learn how to recognise at least the first letters of her children's names so that she could find them in her mobile phone.

In 2018, Neriman had the, very rare for her, opportunity to *gezmek*, or travel, as the *Belediye* was organising a trip for Göreme women to visit Çanakkale and the Gallipoli Peninsula in the west of Turkey. When I visited her the day before the trip, we joked about her becoming a '*turist*' and she got busy packing her bag with long skirts and 'modern' headscarves, which she considered more suitable urban-wear in contrast to the *şalvar* (baggy pants) and *yemeni* (village-style headscarf) she usually wore. Meanwhile, her teenage granddaughter came in and out of her room to check herself in the mirror. Dressed in tight white jeans and a white sleeveless top, she was putting on make-up to get ready to go out to meet friends. Neriman commented, "Doesn't she look like her mother did when she was that age"? I agreed, although Zubeyde, the girl's mother, had not been free to wear jeans or make-up when she was a teenager.

Neriman's daughter was 15 when I first met her during my earlier fieldwork. She had long since left school, and most of the time, she helped her mother do the housework, gardening work, food preparation, and cooking.

When the day's work was done, she and her mother would sit somewhere doing *komşuluk* (neighbouring) with the other mahalle women – often in the area around Hava's house (now Aydınlı Cave-Hotel) – drinking tea and chatting, as well as watching tourists pass by and sometimes making fun of them without their knowledge. Unlike her peers, Zubeyde did not stay married long and returned, with her young daughter – Neriman's granddaughter, to live at the family house in Göreme. Although she had attended primary and secondary school, being quick-witted, Zubeyde had picked up most of her English from me and her brothers' various foreign girlfriends, and this helped her get a job first in a hotel at the reception and later in a hot-air balloon company's office. In doing so, she was among the first Göreme women to work in such 'front-line' tourism positions and, consequently, became the subject of considerable gossip.

It could be said that, because Zubeyde was somewhat ahead of her time in her aspirations to have a life different from her mother's, she did not so much *stretch* the moral boundaries as *step over* them. Within the mahalle and the broader Göreme community, her sense of propriety was deemed 'lost', resulting in often bitter family squirmishes. In conversation, while being seated at the living room of the family house, she told me, in English, so that her mother would not understand, that she had "always felt different"; "Even though I had to be covered (*kapalı*) when I was young, as a teenager I wanted to be free". Referring to her friendship with me, and with the foreigner girlfriends of her brothers, she explained: "I saw another way of living, another way of being a woman; all of you opened my eyes". She told me that, once her daughter went away to university, she too would move elsewhere, perhaps to Izmir or Selçuk; having worked in tourism for several years, she anticipated that it would be easy to find a job. In explanation of why she would leave, referring to Göreme, she said: "I don't like the small-mindedness of the people here, their jealousy and gossip. This town is not for me".

In as much as her own life was quite different from her mother's, Zubeyde hoped that her daughter's life would be different again from her own. She reflected on how quickly things had changed in just three generations: "My mother had no education at all, and now my daughter wants to go to Istanbul University next year to study Law"; "She's smart, and she wants to be a prosecutor", she said, "and after her degree, she wants to do a Masters in Denmark, and then maybe go to the USA".

For those, such as Zubeyde, who were at the forefront of the frontier of change in Göreme, the danger of gossip and chastisement was significant; there have been some instances where the skirmishes ensuing from particular individuals' 'stepping out' were hard to work through. Making it a smooth transition from being *kapalı* to stepping out and crafting new selves seemed necessarily to be a finely balanced frontier, best negotiated in small incremental steps rather than 'leaps'. Nonetheless, the brave leaps of Hanife and Hayriye, with first their stall in the valley and then their modest shop on Göreme's main street, managed somehow to strike the appropriate balance which then allowed other women to feel more comfortable in their own 'space invasions' of the main street. Certain leaps came together to gain their own cultural productiveness so that women were able to find something of a 'place' for themselves, even if a limited place, in tourism work. As one Göremeli woman told me: "Working used to be shameful, but not anymore. Now, to work is not shameful, and to *not* work is shameful"!

Inevitably, then, different age groups growing up with tourism in Göreme have had varying ability to shift moral boundaries by making such steps and leaps of change. Moreover, along with women's repositioning in relation to the tourism economy, other significant changes regarding the life opportunities of girls and women have come about, both enabled and accelerated by broader social fluxes which, somewhat quirkily and unpredictably, became entangled with the tourism-related processes of change occurring at the local level. So, while much of the research on women's work in tourism has emphasised the prescriptive nature of gender norms by focusing on how the norms prevent women from participating equally with men in the tourism workplace (Elmas, 2007; Tucker and Boonabaana, 2012), my intention in this chapter has been to highlight the ongoing potential for change; how transgressions – as well as quirky entanglements – open up new spaces of possibility to rework what is 'doable'. This serves as a reminder, as does the vignette of three generations of women in Neriman's household, that while the frontier of change conjures a spectre for some, even in its quirky unpredictability, the frontier is a space of hope and possibility for others.

Notes

1 I had noted in *Living with Tourism (1)* that: 'Weddings are virtually the only formally organised opportunity for village women to have 'enjoyment' (*eğelenmek*), making them the most conspicuous social events for women of the village' (2003, p. 73).
2 Honour and shame principles have been widely discussed in anthropological literature concerning Middle Eastern and Mediterranean societies (Abu-Lughod, 1986; Bourdieu, 1965; Gilmore, 1987; Goddard, 1989; Peristiany, 1965), sometimes in specific reference to Turkey (Kandiyoti, 1996; Uskul and Cross, 2019) and including in specific relation to gender relations in Turkey (Arin, 2001; Delaney, 1991; Marcus, 1992; Sev'er, 2005; Sev'er and Yurdakul, 2001). As Uskul and Cross (2019, p. 42) point out 'The variety of Turkish terms for the concept of honour (e.g., *onur, namus, şeref, haysiyet, nam, şan, izzet*) attests to its cultural centrality'.
3 This exclusion was not only tourism-related. As Delaney (1991) earlier noted, similar exclusion processes were occurring in Turkey more generally at that time: 'In the process of mechanisation or development, men's focus is being drawn outward; women are being left behind, more than ever enclosed in the house...It is at this juncture...that women tend to become identified with the "domestic" private realm of reproduction and men with the public realm of production' (1991:267–268).
4 See Scott (1995) for discussion of foreign women working in tourism similarly pushing the boundaries of local gender norms in Northern Cyprus.

7 Sticky memories, hopes, and dreams

(T)here's a sweet smell of bonfire in the air, which, when I open the gate to investigate, turns out to be coming from a fire my neighbours have lit right next door and round which they are huddled together with an extraordinary assortment of pots, pans and plastic buckets. I wander over to investigate... "Do you add sugar?" "No... It's all natural" ... As evening draws in, the neighbours are still gathered around the cauldron... Three generations of the family are standing around the fire, feeding vine twigs into the flames. It's an oddly reassuring sight, a reminder that although Göreme often seems to be fast-forwarding into urban modernity, some of the old ways of doing things are still managing to hold their own. Today I doubt that there's any need for people to...make their own *pekmez*... Still, there was always a camaraderie about the protracted preparations for winter, and people seem reluctant to let that go.

(Yale, 2009).

(S)tickiness involves a form of relationality, or a "withness", in which the elements that are "with" get bound together...Some forms of stickiness are about holding things together. Some are about blockages or stopping things moving. When a sign or object becomes sticky it can function to 'block' the movement (of other things or signs) and it can function to bind (other things or signs) together.

(Ahmed, 2014, p. 91)

Having focused in the last chapter on how the frontiers of change shift by way of often unpredictable leaps and skirmishes, in this chapter, I will pay attention to how affect, or 'stickiness', sometimes helps to propel such leaps, and other times enables practices to hold their own, thereby preventing or slowing such leaps of change. To do so, I will use stories, memories, of *pekmez* (grape molasses), as in the above example from Yale (in 2009 a writer and resident foreigner in Göreme's Aydınlı Mahallesi), to discuss how some affective practices and materialities 'stick' and continue to bind people – and people and things –

DOI: 10.4324/9781003011200-7

together. *Pekmez* is indeed sticky stuff, both literally and metaphorically. While it used to be sticky with necessity, now it is sticky-sweet with a blend of habitude, custom, and nostalgia. It is sticky, also, with memories of harder times, perhaps even with a sense of shame associated with those hard times, and yet sometimes that shame folds over and is overrun by anxiety regarding its own loss. So while there is a heavy sense of ambivalence about its stickiness, *pekmez* is, nonetheless, utterly saturated with affect which, for Ahmed (2014), creates 'a relation of "doing"' (p. 91), a relation of moving, which in some instances might be a binding together.[1]

In *Cultural Politics of Emotions*, Ahmed (2014) writes about how objects become sticky 'as an effect of the histories of contact between bodies, objects and signs' (p. 90). Picking up on the historicity of stickiness, in this chapter, I will discuss 'sticky memories', referring to the way memory 'sticks' particular affect to things. In Göreme, this includes the affective dimensions of being 'peasants in transition' (discussed earlier in Chapter 2), or indeed the affective stickiness of the label, or sign, 'peasant' (*köylü*). Whereas in Chapter 2 I considered the particular tourism-moral-economy aspects of the 'new peasantry' (Öztürk et al., 2018), in this chapter, I will focus on what affect does in this milieu. The point that the term 'peasant' (*köylü*), along with *pekmez* and other objects and practices like it, is sticky in the present precisely because of the particular memories that surround it speaks to generational differences regarding tourism development and change since, inevitably, different generations or age groups will vary in their memories, and hence in their relationship to the stickinesses of particular objects, practices, and places. Not only are some sticky memories inevitably stronger for some than for others, but different affective fluxes (Jiménez-Esquinas, 2017, p. 315) may blow in at times to create new regimes of value, thus rearranging old stickinesses to manifest as ambivalence and contradiction with regard to those sticky memories, hopes, and dreams. Hence, conveying reflections on the past, present, and future from young adults, mid-age, and elderly people, this chapter renders a multigenerational and affective perspective with regard to change in Göreme.

Sticky memories

Fieldnote diary, September 2019: I was told to be at Abbas and Senem's house early so that we could head out to the gardens at 7 am. As we loaded crates and buckets into the back of Abbas's car, passing neighbours wished us a good morning followed with "*Allah kolaylık versin*" ("May God make your work easy"), a gesture which acknowledged the 'goodness' of the *pekmez* work we were about to embark on. We headed first to the garden a few kilometres away in the direction of Avanos. In that area I noticed many new houses had been built. Abbas said "Yes, it's becoming our new village out here". "Write it in your book", he said, "that Göreme people are building their house here because they want to escape from the tourists and the noise in Göreme". We picked some grapes in their vineyard in that area, and then moved on to another of their gardens, this one closer to Göreme and in one of the valleys now popular as a balloon take-off area. I asked Abbas why he had put up such a high fence all around this garden – all the villagers' gardens used to be open with no fencing. He said that there were so many tourists passing by here that if they all took just a little of the produce – which they do – there would be nothing left. I told him that when I was hiking near there a week earlier, I had seen a young Turkish couple stripping an apricot tree, and I talked to them asking if the orchard was theirs. They replied that they were visitors, on holiday, and their ATV tour guide had told them they could help themselves to fruit from the gardens. "Exactly", Abbas said, "that's why I built this fence".

There being six of us made for light work: along with Abbas, Senem and myself, Senem's sister and brother-in-law had come from Nevşehir to help with the work, plus Abbas's sister who was back visiting from Holland; she comes each year at *pekmez*-making time. Abbas and Senem's son was unavailable to help since he was at work, selling ice-cream to tourists at the entrance of the Göreme Open-Air Museum. Six of us working together also made for an atmosphere of camaraderie; we chattered away, poking fun at each other for being slack, or slow, or for not being careful enough with the grapes as we cut the bunches off the vines. Sometimes balloons flew overhead, and on the ground, too, many tourists passed by, some on ATVs, some on horse-back, and others on foot. Some of them stopped by the fence to look at what we were doing. Abbas told me to tell them it costs ten lira to watch us, more for a photo; he was joking of course. I said to Abbas that one of the hoteliers in town charges tourists 50Euro to do what I was doing, helping to make *pekmez* for a day. Abbas said that I should give him 20Euro, that would be enough, and later when I dropped a bunch of grapes onto the dusty ground he said that I should pay 30Euro, joking of course.

After four hours of grape-picking (see Figure 2.1) we returned to Göreme and set about creating a *pekmez*-making area in the street in front of Abbas and Senem's house, firstly placing rocks for bollards to block the road to cars. We then set up a 'juicing-station', using a trailer for height and plastic sheeting which was folded up so as to funnel the grape-juice off the trailer and into a large bucket. I got up onto the trailer and 'danced' on the grapes to squish the juice out. While I danced, a *Belediye* (municipality) announcement came through the street's loudspeaker. We understood that the Mayor was asking for all

Figure 7.1 Making *pekmez* in the street in front of Abbas and Senem's house.

wealthy people to donate money for education, although we could not quite hear whether it was for children or women's courses. Abbas said that when he had the tour agency he gave money every year to such causes. "But now I'm poor", he said jokingly, "a poor farmer". I said, "Are you a poor *köylü* (peasant) now?" "No, I'm a çiftçi (farmer)", he said; this time not joking.

As the juicing progressed, a fire was made so that the boiling could begin. A large cauldron was balanced over it, and wood was constantly added to the fire to make the juice boil faster. Each batch needs to boil for several hours to render it down to the thick *pekmez* molasses. A Japanese tourist came by and stopped to look at the various activities going on – Senem and her sister looking after the fire and stirring the cauldron's contents, myself still squishing the grapes on the trailer, and Abbas and his brother-in-law working to collect the juice I made in large buckets. The tourist appeared particularly interested in the fire and the two traditionally clad women working at it, and he held up his camera to ask if he could take a photo. Senem's sister shouted, "Pay $10!" at him. When he started to take the money out from his pocket, however, Abbas and I quickly explained that the women were only joking and that, yes, he could take a photo. To make amends after her joke had been taken too seriously, Senem's sister offered the man a small bowl of *pekmez* to taste, which he accepted with apparent delight.

A while later, some neighbours came by and put pots of *güveç* (stew) into the fire's embers to cook. It was time for us to take a break by then, and Senem placed aubergines and peppers into the fire so that, when they were amply roasted, we sat on crates to eat our street picnic of the roasted vegetables stuffed into bread. Abbas and Senem's niece, Senem's sister's daughter, who had recently married a Göremeli man, came by and joined us for the picnic and helped to stoke the fire for a while. The group chatted about how some people sell their *pekmez* for 35 tl/kilo, with that being all they can charge because they do not reduce it down enough, leaving it runny. Abbas said he will sell for 40 tl/kilo this year because they make theirs very thick.

Figure 7.2 Neighbours' stewpots and peppers roasting on *pekmez*-fire embers.

The *pekmez* work continued as the afternoon turned into evening. Various members of our group took turns in periodically going into the courtyard to pray on the prayer-mat Senem had earlier laid out. Anybody who came by, and who said "*Kolay gelsin*" (may your work be easy), was offered a small bowlful of *pekmez* fresh from the cauldron. In the late afternoon, another niece, the daughter of one of Abbas's brothers, stopped by with her two young sons. She sat and chatted with me for a while, and when she got up to leave, Abbas said, with disappointment in his voice, "Are you leaving already?" A while later, Abbas had a call from a friend asking if he was coming to the teahouse. He said that he couldn't, he was making *pekmez*, and that he had put potatoes in the fire to bake and his friends should come over to help eat them. They did not come, but instead Emre came home from work and he helped eat the potatoes. I heard Abbas telling Emre that his cousin had come by earlier with her sons but they had stayed for less than half an hour; "How can the children learn if they only stay for half an hour?", Abbas lamented. Emre did not stay for long either; he soon went off with the tractor to deliver tanks of water to hotels at the top end of Aydınlı Mahallesi, where the water had cut off again.

They say that people have been killed by *pekmez*, when the cauldron tipped off the fire and spilled out its burning sugary stickiness. They also say that before the arrival of beet sugar, *pekmez* was the only way to sweeten tea, and so that is why *pekmez* used to be sticky with necessity, or – perhaps we can say – necessarily sticky. Besides, during Cappadocia's cold snowy winter months, *pekmez*'s energy-giving properties have always been a lauded source of health and wellbeing. Abbas and Senem would never dream of going through winter without their own *pekmez*, even though the actual necessity has long gone. The necessity of *pekmez* and of other food production activities was altogether more present in people's memory during my earlier fieldwork, conducted in the 1990s. By then, the income from tourism had rendered it possible to purchase most food items from the market and so the food production activities which had previously been an integral part of Göremeli people's lives had already declined in importance. Yet, their necessity seemed to have stuck. As I wrote in *Living with Tourism (1)*:

The memory still lives on of times when survival of the household depended directly on working hard in the gardens, and it seems certain that such memories of hard winters and empty stomachs will die hard, even though those times now seem to be firmly in Göreme's past.

(Tucker, 2003, pp. 84–85)

I wrote also that Abbas's memories 'of his childhood and of the difficulties his father had in maintaining the household are always present and underlying his (and those of others like him) experiences today' (ibid., p. 74). As such, the sticky memories of how life used to be continued on as reminders, creating an ambivalence once the earlier necessity of the hard work was gone. This ambivalence was illustrated in an argument between Abbas and Senem I overheard during my 1990s fieldwork where Abbas wanted to "burn the field" of wheat because it was easier and cheaper to buy flour rather than hire a thresher and grinder to process the wheat themselves. Senem pronounced that "it is *adet*, we are used to doing it", and then Abbas's elderly mother added: "Before, doing the wheat was the only way we could eat". In the 1990s, I regularly heard the older generation reminding their children and grandchildren that they must work hard because otherwise they would not have bread, or tea money, for the winter; time still passed in a cyclical manner, with each year being about producing enough food as well as enough income from tourism during the summer months in order to be able to survive through the coming winter.

As already explained in earlier chapters, at the time of my 1990s fieldwork, the lives of villagers, especially women, were still filled with agricultural and food production activity, even if by then, it was mainly for domestic consumption. Every morning throughout the spring, summer, and autumn, women went off to their gardens on donkeys or horse-drawn cart to tend to the soil, prune vines and trees, and harvest produce. Back at home in their mahalle (neighbourhood), elaborate processes of preparing, drying, and preserving food were carried out. Around the communal ovens dotted around the mahalles, usually set into a stone wall or the rock-face of a fairy chimney, the women would get together to bake pastries and breads stuffed with potato and onion, or to roast pumpkin seeds. Groups of neighbours often came together in this way, forming a *topluluk* (community) to share the larger workloads involved in making bread, or macaroni, or *pekmez*. For the production of flat-bread, groups of women would gather together over several days to make a large pile – six months' worth – of the paper-thin bread for each household of the group. I wrote: 'As they work, they joke and gossip, and are able in this way to keep up the hard work from early morning to well into the evening if necessary to finish the task' (Tucker, 2003, p. 83). This co-operation, especially between women, was not only a fundamental part of mahalle life, but it also exemplified the peasant sociality of balance and necessity (discussed in Chapter 2).

Now, 20 years on, perhaps with the exception of *pekmez*, it tends only to be the less arduous tasks which are still undertaken. While a few households continue to make their own bread and macaroni, they no longer grow the wheat to make the flour, and most families either purchase the processed options from the supermarket or they source homemade goods made by women from other Cappadocia villages where the tourism economy is not so well-established. During the past two decades, I have no longer heard stories told of the earlier hard times, or of sticky necessity. Not only has the hard necessity

Figure 7.3 Depicting *pekmez*-making as Cappadocia's cultural heritage in an Avanos museum.

of homebased food production dissipated, but so too to a large extent, has its memory. *Pekmez*-making, and other similar food production activities such as *acılı* (hot pepper sauce) and *salça* (tomato or red pepper puree), are now sticky with the sweeter memories of camaraderie and nostalgia. Indeed, for Abbas and Senem, the stickiness remaining is that which today lingers as *adet* (custom), and, as Yale (2009) quoted in this chapter's opening extract said, it is the joy of the camaraderie that people seem reluctant to let go of; a cooking fire in the street still becomes a focal point for family and neighbourly togetherness. Yet, these days only the occasional cooking fire can be seen in the old mahalles, where the streets are instead filled with the bustle of tourist rental cars, delivery vans, and airport shuttle buses, heading to and from hotels. In the Yeni (New) Mahalle, where Abbas and Senem live and which still functions as a 'neighbourhood', while such food production activities continue to be prioritised enough to warrant closing a section of road, it is mainly the older members of households for whom the sticky memories are strong enough to consider it important to continue.

Some have tried to relegate *pekmez*'s stickiness to the past; in ethnology museums in the nearby towns of Ortahisar and Avanos, *pekmez*-making – as an aspect of Cappadocia cultural tradition – is described using the past tense, as if it no longer continues as a practice in the present day. While the museums render *pekmez* sticky under certain heritage regimes of value (Jiménez-Esquinas, 2017), it is this process of 'heritagisation' which also relegates it to the past. For many young Göremeli, too, *pekmez*-making is a thing of the past – practised by their grandparents and perhaps by their parents – and not something they themselves are interested in; they are happy for *pekmez* to belong to the museums from now on. This is where tensions arise, however. Abbas appeared sad that his niece, Sonbahar, who came by with her children, would not engage more: "How can the children learn if they only stay for half an hour?"; how can the sweet sticky memories rub off on them if they stay for such a short time?

A few days after that *pekmez*-making day, Sonbahar's sister Leyla arrived home in the village and I caught up with her when she was hanging around a neighbour's *pekmez* fire helping out. She had returned from India where she was living with other like-minded travellers, and was pleased to have timed her return visit during *pekmez*-making season. Over coffee a few days later, I asked both sisters if they knew how to make *pekmez*. Sonbahar responded that because their aunties had made *pekmez* since they were little girls, she had a rough idea of what goes into it; "But I don't really know how to make it", she said in a way which suggested that my question had prompted her realisation of this. Now in her early thirties, Sonbahar went on to say how she had had a lucky break from the necessities of things, such as learning how to make *pekmez*, since she was among the very first Göreme girls to go to high school: "I was lucky because for my age group it was more normal for girls to get married and not go to high school". The two sisters were, it could be said, a 'product' of tourism, since their British mother had been among the first 'tourist brides' to marry a Göreme man (Abbas's brother), and they grew up in the backpacker *pansiyon* their parents owned and ran. In addition, their British mother and grandmother were keen that the girls should get a good education. Being seven years younger, Leyla was afforded even more educational opportunities than her sister. She was sent to private schools in Nevşehir from the start of her schooling, and went on to win a prestigious scholarship to attend university in Istanbul. During the university summer breaks, she went to stay with her grandmother in Scotland and worked in an Edinburgh bar, an occupation far away from *pekmez*-making.

Perhaps, it was this distance from sticky necessity that later enabled Leyla to return home from her travels with a fresh interest in the traditional food practices of her aunties and uncles:

> While I was growing up, the whole thing started about getting educated and having a better future, and all of my generation forgot about what happens here. At that time, I wasn't interested at all in farming and all these things, but I'm really interested in it now... I really want to get out of the system of buying everything, and thought the best way to do it is to grow my own food. But even though my whole family are farmers, I realised I had no idea how to plant anything. That's because I didn't see

it as a fun thing, because whenever we went to the fields with the family, it was just a job that needed to be done. And some of my friends did go and work in the fields, so they learned how to do it, but now they're not doing it themselves, because they moved to the city and got married, and they're like "Why should we?"; they get a job and can pay for their food, and they don't find it interesting to make the food themselves…

So these days, whenever I'm back, I ask my aunties about different things, and I'm like 'Wow – no one knows this anymore'! So this is my big fear, that all of this is going to get lost… that none of this is going to still be here – because it is knowledge you can't learn from books, like how to prepare the different foods – there's these little things they have to tell you – you can't go on Google and search it – and so this is my really big fear.

Along with some sticky memories from her childhood, for Leyla these days, much of the stickinesses related to local knowledge and practices of food preparations might be connected with broader trends such as the global slow and sustainable food movements. Such movements, for instance those identified by Kathleen Adams (2016) as part of an 'Eat Pray Love' trend in global tourism, act as what Jiménez-Esquinas (2017) terms global 'affective fluxes', which are 'forces' that 'do things' by 'imposing their sense of what is valuable and important and what current generations must care about' (ibid., p. 315). In other words, the food practices which Leyla grew up with, but in which she used to have no particular interest, are now assigned new value, or new stickiness. For Abbas and Senem, on the other hand, *pekmez* bears a visceral stickiness of belonging, or perhaps, in Ahmed's (2014) words, of binding. For them, they only dare to hope that their son will marry soon, and that he and his *gelin* (bride) will live downstairs in the house they have prepared for him on the bottom floor of their own house; keeping them close would hopefully mean that their daughter-in-law and grandchildren would come to appreciate *pekmez*'s stickiness, which Abbas and Senem themselves are reluctant to let go.

It is interesting to wonder what relation or convergence there might be between the affective valuing of *pekmez* for Abbas and Senem and that for Leyla, and Sonbahar for that matter. Indeed, Sonbahar agreed with her sister that "It's not going to be with us very long, *pekmez*, I feel". However, this was said in a rather neutral tone compared with Leyla's somewhat impassioned plea for it all not to be lost. So is it, then, that Leyla's fear that "all of this is going to be lost" is derived much more so from regimes of value produced by global affective fluxes than from the memories of a young girl who grew up in Göreme? It is also important to consider what part such global fluxes and their associated regimes of value play in my own affective relationships and writings; what particular regimes of value render *pekmez* sticky for me? Furthermore, where do the global fluxes which rendered it possible for women of Sonbahar and Leyla's generation "to get educated and have a better future" fit into all of this?; what of the part played by the global feminist movement, and the education reforms discussed in the last chapter, in enabling many of the old stickinesses to come unstuck? There are inevitable contradictions and ambivalences when it comes to memories, regimes of value, fears, and hopes.

Hopes (of becoming other)

Fieldnote diary, October 2009: We were told to be there early, as we had a big day ahead of us. We knocked on 'Neriman-Mother's' door at exactly 7 am and when she opened it she looked surprised, as if she hadn't expected us to actually show up; as if she had been afraid to get her hopes up too much. A few days earlier, we'd been chatting with her in her son Mehmet's café, where she helped out sometimes making *gözleme* (a stuffed 'pancake'-type food) in the traditional way over a fire in the café's front courtyard, giving the café an additional attraction for tourists as well as an extra menu item. While chatting, I'd asked her if she'd made *pekmez* yet this year. She said that she couldn't because she had no one to help her; Mehmet and his sister said they were too busy in their tourism jobs, and another of her sons living in Avanos was also busy working. Her other two sons lived in Australia at that time. She said that all of her children had told her not to bother with the gardens anymore; it's too much work for her, they said, especially since her knee replacement operation earlier in the year. However, it was because of her knee problems that she hadn't been able to work in the gardens last year either. For over two years, she hadn't even been to her fields to see the grapes, let alone do anything with them. Then she asked whether we would use our car to help: Would we drive her to the field and bring her back with whatever grapes she managed to find worth picking? And so the arrangement began whereby we would spend the day helping her and in return she would teach us the whole process of *pekmez*-making.

A little after 7am, having packed the car with buckets and crates, we drove out to the field a few kilometres from the village in the direction of Avanos (the same area where, ten years later I drove with Abbas and Senem and noticed many new houses). When we first arrived at the field, Neriman was worried; the vines looked dry and swamped by weeds. However, on closer inspection, there were lovely grapes and lots of them. We set to work and as we picked and placed bunches of grapes into buckets, Neriman talked about how she had thought she would have to give up on her vineyards and sell them. Now, though, she wouldn't; she would come back in early winter to tend the vines, so that she'd definitely be able to make more *pekmez* in the years to come. As we loaded the car with all the crates and buckets full of grapes, Neriman said she would meet us at the road. Since she was unsure when she would be able to get to the field again, she wanted to walk across her land once more before we returned to the village.

Arriving back at the house in Aydınlı Mahallesi, we unloaded the car taking the crates of grapes into the back courtyard. Neriman went to the storage cave underneath the house and brought out lots of things; huge cauldrons, a tyre, buckets and scoops. A large sack was filled with grapes, and a special powdery '*toprak*' (soil) – apparently dug from a mountain near Gülşehir – was added to the sack (we learned that this *toprak* would act as 'fines' to render the *pekmez* clear when it was later let to rest after its first boiling). We washed our feet and climbed onto the sack full of grapes. Trying to keep our balance, we started dancing and, as the warm pink juice came out, it bathed our bare feet in a lovely softness. As the juice sloshed around, Neriman scooped it into a cauldron. Neriman's son Mehmet emerged from his room with his baby daughter and he made us coffee. There was a convivial atmosphere among us all and Neriman was clearly very excited; perhaps almost too excited, as Mehmet kept telling her to calm down. A short while later, as Mehmet went across the courtyard to go out to his café to work, he shouted 'Hazel and Rob, thank you very much. You've made my mum very happy'.

Figure 7.4 Our *pekmez* work in Neriman's backyard.

Soon, Neriman lit a fire between some bricks. While the first lot of juice was heating up we went inside and had eggs, bread, and tea. As we ate, Neriman reminisced about the time of my earlier fieldwork visits, about how we all sat together in the street in a big group of neighbourhood women and girls – everybody crocheting and everyone happy. Neriman's daughter had arrived home from her early morning job in a balloon company office just in time to join us for the breakfast. While eating, I said to her, "Don't you do *pekmez* anymore?" – "No", she said, "not anymore".

Figure 7.5 Neriman making *pekmez*.

As the afternoon progressed, we took turns in being the grape-crushing 'dancer'. The juice we produced was heated in batches, then let to sit and cool, and then heated again, this second time having to boil rapidly for a couple of hours or more. All the while we scooped out the froth that formed on the top of the bubbling molasses, and stirred to prevent it from burning at the edges. When the first batch was becoming thick, Neriman scooped a little out onto a saucer. After waiting a short while to see if it became the required sticky consistency as it cooled, she offered it to me saying, "You taste it".

The *pekmez*-making continued on into the evening, and when we decided it was time to leave, Neriman, who was almost bent double by then because her back ached so much from the work, said repeatedly, "May God bless you all, may God bring goodness to you". She said that she would not sell her fields now. Instead she would always find a way to go there, to tend them.

Neriman did carry on – tending her gardens, harvesting the produce, and making *pekmez* – in the years that followed. Sometimes Mehmet helped her, and when another of her sons returned from Australia, he and his Australian wife helped too. Neriman also said that my partner Rob was like another son after helping her so much – we had helped her salvage her grapes and vines, enabling her to continue being a *pekmezci* (*pekmez*-maker), she said – so since the time of being *pekmez*-makers together we have referred to her as 'Mother'. Mehmet told us when we were at his café a few days after the *pekmez*-making day, however, that although we had clearly made his mum happy, he did not really want to see his mum still doing such things. He was of the view that *pekmez* – along with homemade bulgur wheat and dried grapes – should be left in the past; "We should bulldoze the vines", he said, and went onto explain that the Göreme people should move on and no longer be engaging in the 'peasant' activities of before. In any case, he said, he would rather his mother kept her time free for helping in the café, or for looking after his baby daughter while he and his partner worked. So whilst, as I know now, Neriman did continue being a *pekmezci* throughout the 2010s, her being so was always shrouded in ambivalence, between the tug of the past and the pull towards a different future.

I had spoken to others that same year, 2009, who conveyed sentiments similar to Mehmet's regarding their *köylü* ('peasant') past. For example, another businessman, Faruk – then in his mid-twenties and a few years younger than Mehmet – told me that when he was young, all of his family survived by going to the garden; "But it was a really hard job", he said, "tourism is also hard but it's a much better life than farming". Reflecting back on Abbas, then, it is understandable that his return to farming is, too, met with ambivalence.

On the one hand, he is admired for his return to living off the land; on the other hand, the admiration is often conveyed with a smattering of derisive incredulity; while it is hard work to farm, and it has always made for a precarious living on its own, to be *hardworking* is deemed honourable, as is to live off the land. It is therefore difficult to let go completely, even if "tourism is a much better life." So, whilst agricultural and food-preparation practices have lessened considerably in Göreme over the past four decades, with many people either selling their gardens, vineyards, and orchards or not tending them so that they become unproductive, many people have held onto their inherited plots of land. In line with Öztürk et al.'s (2018) suggestion that the contemporary, or new, 'peasantry' in Turkey 'continue to value their land and farms and communities for their own sake, as ends in themselves' (p. 251), many of Abbas's friends, even whilst chasing wealth from tourism, have tended to hold on to at least one or two of their gardens. Indeed, it is their tourism wealth which *enables* them to choose whether or not to hold onto their gardens, just as it is Abbas's previous entrepreneurial ventures which enable him, now, to live comfortably while returning to farming.[2] After working for almost three decades doing tourism business, Abbas and Senem's return to living off the land has involved using their produce largely for household consumption, while also selling a small amount of excess foodstuff, including *pekmez*, roasted pumpkin seeds, and 'çömlek' (pot) cheese, to fellow villagers. In productive years, they also sell grapes via a road-side stall or to a nearby winery. Yet, Abbas, and his friends, appeared reluctant to accept whatever stickiness comes with the label of '*köylü*' (peasant), a label which, through time and in different cultural contexts, has presented 'an undeniable combination of opposite elements' (van der Ploeg, 2008, p. xiii). Along with the pride and dignity that comes with relative independence, self-sufficiency, hard work, and moral integrity, *köylülük* ('peasantry') in Turkey, as elsewhere, has long held associations of 'backwardness', thus always seeming to incite a sense of shame. Abbas clearly rejects any such sticky associations, and declares that, since retiring from tourism, he has *become* something else, something new, a farmer (çiftçi).

Indeed, the desire to become something other than *köylü* has intensified throughout the four decades of Göreme's tourism. While such aspirations were likely initiated several decades ago, prompted by the opening to a market economy and outward migration programmes that occurred in the 1950s and 1960s, during the early years of tourism development in the 1980s and 1990s the ways and possibilities of becoming 'something other' multiplied. As I wrote in *Living with Tourism (1)*, the 'romantic developments' between Göreme men and tourist women were seen as a possible route to a prosperous future (Tucker, 2003). While, on one hand, the 'tourist girlfriends', or 'brides', often invested money, or at least their labour and knowledge of tourist markets, into their partners' businesses, these relationships also boosted hopes of becoming something other and better by easing the way for Göreme men to get a visa to live in another country, which invariably was seen as a land of prosperity. Over the years, the numerous marriages between Göreme men and 'tourist brides' led to Göreme men living abroad – in England, Australia, New Zealand, and so on. Many such relationships led to disillusionment, however, with marriages breaking down, plus the realisation that gaining employment, let alone prosperity, in overseas countries, was not as easy as previously believed. Concurrently, tourism developments prompted a growing realisation that it may not be necessary to leave Göreme in order to achieve prosperity. For the first time, the people of Göreme were able to see the possibility of making good *within* the village, thus lessening the urge to live and work elsewhere.

It was not until tourism was really taking off in the 2000s – with tourists and tourism business opportunities becoming much more plentiful – that people really dared to hope that they could make a better life for themselves and their families *in* Göreme. As Anderson (2006) argues, hope arises through particular intersecting occurrences 'dimly outlining the contours of *something better* and therefore enacting potentialities and possibilities' (p. 749). This outlining of the contours of something better is likely how 'a space of desire' which 'once glimpsed, one cannot but explore' (Tsing, 2005, p. 32) is created. Along with tourism business and education reforms creating spaces of desire, in Göreme, the contours of something better have inevitably been outlined by seeing how 'others' – be they visiting tourists, resident foreigners, or families on television – live. Relevant here is Soares' (1998) discussion of cultural mimicry, whereby globalisation processes increase 'the opportunities for copying by performing one's own fantasies about the other' (p. 295).[3] An example is the Göremeli men who became hot-air balloon pilots after observing the many foreign pilots who had come to work in Cappadocia during the 2000s and who, since being a balloon pilot held significant prestige, often had an air of superiority about them. Many young men of Göreme who might previously have seen moving to a foreign country as the best path to status and upward mobility consequently endeavoured to train as balloon pilots and thereby to gain status within their own village. Emulating their foreigner counterparts, local men working as pilots would walk around in their 'pilot uniform' (lapelled shirt and designer sunglasses), which appeared to afford them significant status among their peers as well as success in wooing tourist girls in Göreme's bars. Soares (1998) suggests that in such 'mimicry', the 'experience of being something else… brings the possibility of circulating, shifting, and changing to the forefront of social and cultural life' (p. 295).

The notion of mimicry might be linked also to the propensity for tourism encounters to create a mirror effect whereby, in each party's encountering of the other, they see themselves as they *imagine* they appear in the eyes of the other. In Göreme's tourism context, these imaginings may very well be sticky in their conjuring of self-associations with *köylülük* (peasantry) 'backwardness', hence inciting a sense of inferiority and even shame. In the earlier days of Göreme's tourism when the majority of tourists were Western backpackers, however, this inferiority-inducing mirror effect did not seem so prevalent. While some elements in Turkey of valorising of the European 'West' are undeniable, as I wrote in *Living with Tourism (1)*, there was a kind of levelling whereby Göreme hosts demanded a relationship of mutual respect with their tourist guests. In contrast to this, when domestic, or Turkish, tourists started visiting Göreme – many of whom came from the main cities such as Istanbul, Ankara, and Antalya – an inferiority complex took hold. Along with an apparent division in Turkish society between an urban educated elite and 'country folk', or peasants, with many of the visitors being of a 'new rich' set, a kind of class differential was clearly felt. I often heard Göremeli hoteliers and other villagers saying they "do not like *yerli* (domestic) tourists", accusing them of being snobbish, arrogant, and fussy, and when I observed even the smallest service encounter between Turkish visitors and Göremeli, I discerned a sense of discomfort on the part of the latter. With Göremeli people keen to shed the identity of the backward *köylülük* (peasantry), it is little wonder that Abbas was always quick to correct my suggestion that he reverted to being a *köylü*, stating instead that he has become a farmer (*çarşı*).

The 'open' and 'closed' metaphor, too, is significant in relation to the desire to become something other than *köylülük*. As discussed in Chapter 6, village life – and particularly that of Göreme women – has long had associations of *kapalı* (enclosed or covered), but an open-closed binary is often used in referring to people's mindset also. For example, Faruk, quoted above, reflected on how tourism had allowed the Göreme people to become "more socialised, more open-minded; because of tourism we meet people from all over the world". In other words, through tourism, Göremeli people, entrepreneurs in particular, gained what Natalia Bloch (2021) describes as a 'familiarity with the global world and high level of intercultural competence' (p. 115). Notar (2008) similarly refers to such an 'openness to the world' being developed by 'locals' who work in tourism, suggesting that the initial binary opposition established by Hannerz (1990) – between 'cosmopolitans' who travel and have a worldly outlook, and 'locals' who do not 'see' beyond their own community – is challenged by tourism's ability to create cosmopolitanism among 'locals' without their necessarily going anywhere at all. Indeed, in their dealings with tourists, Göreme 'locals' – the men at least – who work in tourism evoke the notion of cosmopolitanism in their 'personal ability to make one's way into other cultures, through listening, looking, intuiting and reflecting' (Hannerz, 1990, p. 239), as well as a 'built-up skill of manoeuvring through systems of meanings' (Vertovec and Cohen, 2002, p. 13; see also, Tucker, 2011).

This skill was illustrated clearly to me one day when I visited my friend Fatih's carpet shop. At the time two elderly Afghani men from the Grand Bazaar in Istanbul, both with long white beards and in Afghani dress, were visiting my friend's shop to try to sell their wares. Fatih guessed that I might want a photo with the interestingly apparelled Afghani men and so he set about introducing us in such a way that they and I felt enough at ease for a photoshoot to ensue. Just as he had previously succeeded in manoeuvring through tourist systems of meanings related, for instance, to how he might adapt his shop to cater to the growth of tourist *Instagram* photography, my friend was able – rather successfully – to mediate and negotiate multiple systems of meaning in the presence of myself and the Afghani men; his 'skill of manoeuvring through systems of meanings' was built-up through prolonged living and working with tourism. As I wrote in my earlier work (Tucker, 2011, p. 34), the people of Göreme (again, mainly men) who have been working in tourism for multiple decades now have not only themselves had the experience of repeated and prolonged interactions with tourists, but also the chance to observe each other interacting with tourists and thereby to share and exchange their tourism know-how. To refer to the words of Faruk again, he and others like him have become more *open* to the world because of tourism; they have met their own desires to become something other than *köylü*.

As discussed in Chapter 6, these cosmopolitan skills and the ability to become something other than *köylü* were not necessarily afforded to Göreme women as they were for men. Having remained relatively isolated in the 'domestic' space of their immediate mahalle, those women who make inroads to finding a 'place' for themselves in the tourism business were less able to acquire the necessary skills and knowledge. In other words, women's hopes of becoming 'something other' were impeded by moral imperatives for women to be *kapalı*, so that their experience of 'crafting new selves' was inevitably an ambivalent one. In particular, the older generations' sticky memories – and corresponding chastisements and skirmishes discussed in Chapter 6 – tended always to tug back on the desires and hopes of the younger generations. Yet, both Göreme's tourism and the education reforms enacted across Turkey – in particular, the remarkable increase in access to higher and tertiary education for girls and young women – afforded significant new 'potentialities and possibilities' during the 2000s and 2010s. As early as the mid-2000s, I recorded in my fieldnotes a woman who worked in a hotel telling me:

> The way people think has changed. Young people have changed. The phrase '*Evde otur*' (Sit at home) is finished for young people, completely finished... Education is really important here now; even the girls go to school. Before they used to learn only how to work in the fields. They do not learn that now. Now they only want to study.

This sense of young people thinking differently from the older generations was similarly expressed by Faruk, quoted above, when he continued to say regarding the matter of open-mindedness:

> I'm already more open-minded than my elders, and I hope the young people become even more so. We need to stop being narrow-minded. The older generation will definitely say that people are losing their culture because of tourism, but my generation wants this change.

This hope that younger people will become *even* more open-minded was seen, too, in the vignette in Chapter 6 of three generations of women in one household. For Zubeyde, while 'the weight of the sedimented past' (Puwar, 2004, p. 1) was so overbearing as to limit her own possibilities to be "different", "to be free", she had big hopes for her daughter to have a different life from her own. Such hope was also illustrated in a conversation I had with Mustafa, the successful restaurateur introduced in Chapter 2, who told me that he had sent both of his daughters to university so they could have an "open future": "They can take over my restaurant, or not, they choose. I want them to have an open future", he said.

This idea of an open future resonates with Phoebe Everingham's (2016) suggestion that hope exists in '*the-not-yet-become*' (p. 522). Indeed, such hope appeared to be the mood more broadly in Göreme towards the end of the 2000s, at the time of the *pekmez* story which began this 'Hopes of becoming other' section; my fieldnotes from the circa 2009 period are full of hopeful sentiments. For the first time, I was hearing the aspirations of young women to become nurses, lawyers, or accountants, and these aspirations were, without doubt, hopes of something better, of '*the-not-yet-become*'. Moreover, there was a strong sense of hope more generally regarding tourism and what tourism business could bring. As discussed in previous chapters, a lot of people were converting cave-houses in the old mahalles into hotels at that time, and those who already owned a hotel endeavoured to upgrade and expand their business. During that period, there was a significant proliferation of businesses, and with those businesses came the anticipation and hope of becoming something *other*, something *better*.

Among the many examples of hope-filled business development were those undertaken by Osman (introduced in Chapter 2) who, having been one of the first Göreme men to develop a tourism business back in the 1980s, bought a large plot of land at the top of Aydınlı Mahalle in the hope that, one day, he would develop it. In 2019, after several years of stressful and costly construction work, Osman finally opened his new high-end 'boutique' cave-hotel. Set around extensive grounds with grass lawns and a large central reception room and restaurant, the hotel's cave-suites, many with private terraces, were fitted with delux but quirky opulence such as circular beds and, in a 'family suite', a mini toilet and wash-basin for young children. As Osman showed me around his hotel, he told me that owning such a place had always been his dream. From starting out selling postcards to tourists as a teenager, to opening a tour agency and small hotel, "this hotel is the end result", he said, "the end product"; "This hotel is my dream!"

Dreams

Fieldnote diary, October 2019: I was told to be at the hotel early, since the excursion to the valley for breakfast and *pekmez*-making would leave at 8.30 am. When I arrived, the transport — a large trailer pulled by a tractor — was already filled with hotel customers of various nationalities:

Irish, New Zealander, Indonesian, American, Saudi, Indian and English. We set off and once we had climbed up the bumpy and dusty road leading out of the village from the top end of Aydınlı Mahallesi, the tractor stopped at a point with good panoramic views so that we could take photos and have an introductory talk: "I'm Ali, the owner of the hotel and I am your host. I'm going to tell you some things about this place. We used to come up here on a donkey, every family had a donkey and if you were rich you also had a horse. Life was hard and everybody lived from their gardens and fields. In this area we had lots of wheat fields. It was very hard work harvesting the wheat and threshing it using a donkey. We would make bulgur wheat and keep some aside in a dry store cave in case of a drought year; we were very poor then. In the 1960s and 1970s many people went to Europe to try to earn money. Some of them are still there and some came back". He then explained that tourism had started in Göreme in the 1980s, and he had opened his hotel by converting his family's cave house in 1993. Most of his brothers had gone to Europe but he had stayed and opened a hotel. To this, a man from Ireland commented that Ali was "the smart one".

We travelled further up the dusty track until the point where we would leave our transport to walk down the steps cut into the rock to the bottom of the ravine. There − at Ali's 'garden place' − a magnificent breakfast table was laid out and Ali assured us that everything was grown and made in the valley, "Probably even the President doesn't eat such a good breakfast", he said; "It is all organic, using pigeon droppings for fertiliser". While we ate, several women worked behind us in the nearby outdoor 'kitchen', continuing to prepare food to bring to the table, and serving us teas and coffees as required.

After breakfast, we moved to the other side of the ravine floor where an open-fronted stone arch-room was set up to display a 'traditional' cave-house and way of life, with a '*tandır*' oven in the dirt floor to show how cooking used to be done. Ali did another talk, this time explaining the relevance of the cave houses for keeping cool in the summer and warm in the winter; "If you didn't live in a cave house, you were miserable in Central Anatolia". He talked of keeping grapes, apples, and pears fresh throughout the winters in the damp cave downstairs, while the upstairs cave rooms were for living in and for storing dried goods, such as dried grapes, apricots, walnuts, and pumpkin seeds. Walking further up the ravine, Ali pointed up high above us to focus our attention on the many 'pigeon houses' carved into the rock-face. He explained how important pigeons had always been in Cappadocia: "The droppings were collected for fertiliser, so our crops were organic; pigeons were so important that you couldn't find a girl to marry if you didn't own a pigeon house". Ali then explained that when people started working in tourism,

Figure 7.6 Ali explaining how Cappadocia life used to be.

they stopped tending the pigeon houses and so the pigeons went away. He said that one of his recent projects was to restore pigeon houses and to bring pigeons back to the area in order to keep the tradition going because of its importance for health: "People were healthy when their food was organic, they didn't need to go to the doctor. Now they use chemical fertiliser and they get sick".

It was then time for most of the tourist group to return to Göreme, but myself and a couple from Indonesia, plus the couple's tour guide, were taken in a 4×4 vehicle up out of the ravine to a field where we would begin the '*Pekmez tour*'. The couple had purchased this *pekmez*-making experience from Ali's travel agency business, from where they also had hired the guide to construct a personalised itinerary and take them around while in Cappadocia. Ali had invited me to join the breakfast and the *pekmez* tour because he knew of my interest in the relationship between tourism and local culture.

At the garden, we were taught by Abdullah, a staff-member working for Ali, how to cut bunches of grapes off the vines. We worked to fill tall wicker baskets and were glad that Abdullah and another worker, a younger man from Afghanistan, were there to carry the heavy filled

Figure 7.7 Tourists picking grapes during the '*pekmez tour*'.

baskets to the jeep. When Abdullah decided we had enough grapes, we returned to the breakfast ravine, where the *pekmez*-making was to take place.

Some young Afghani workers filled large sacks with the grapes and placed them up on a roof where we would do the stomping to juice them. For that job, we were each given a pair of new yellow gumboots – still in their wrapper – and after putting them on, the Indonesian couple, their guide, and I took turns dancing on the grapes. This activity created much fun and plenty of photos and videos were taken, some of which the Indonesian couple immediately posted on social media. The juice was funnelled down off the roof and caught in a cauldron below. When there was enough juice, one of the women who had earlier made our breakfast lit a fire and started to heat the juice-filled cauldron. We went down and joined her and each of us were given a go of stirring the big pot of boiling juice, providing more good content for social media-worthy photography.

It took the remainder of the morning to complete the *pekmez*-making process, at the end of which we were each given a bottle of the sticky molasses to take away with us. The remainder of the *pekmez* we had

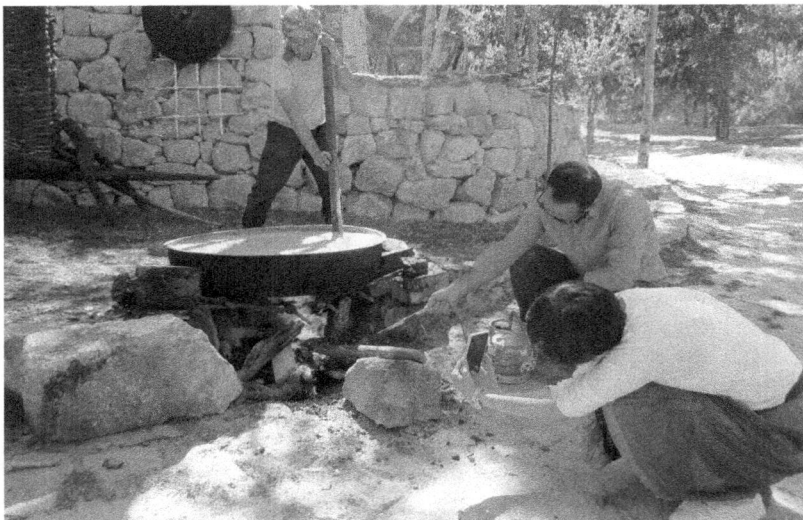

Figure 7.8 Tourists and guide enjoying making pekmez during the '*pekmez tour*'.

helped make would be used in future tourist breakfasts in the ravine, and it would feature on the breakfast buffet in Ali's hotels. We had only picked and used a small portion of the grapes grown in Ali's many gardens, ensuring that plenty remained throughout the *pekmez* season for any other tourists who may wish to book the '*pekmez tour*' and become *pekmez*-makers for a day.

It is interesting to consider what moved Ali to create the kind of tourist activities and experiences portrayed above. In 2019, when I joined this breakfast and *pekmez tour*, Ali's business portfolio – five hotels, two hot-air balloon companies, a travel agency business, and more – was such that he did not have a business need to do so. With approximately 250 employees, he had no particular need for face-to-face dealings with his tourist customers at all, and yet he acted as 'host' during the breakfast and told his stories about Göreme's past times and the importance of pigeons. His motivations were made somewhat apparent when, upon realising at the start of the 'breakfast-tour' that some of that day's guests had no understanding of English whatsoever, Ali appeared deeply disappointed. The breakfast, it seemed, was a means to gather his hotels' guests together as a captive audience for his stories; through telling his stories he hoped to inculcate tourists with an interest and respect for Göreme's history and culture.

Moreover, I learned through talking with Ali that another, perhaps greater, purpose of these tourist activities was to create a show of using and valuing the valley gardens, of working to keep the gardens productive, and a show of reaping "organic" – and therefore healthful – produce and local traditional foodstuffs to serve to tourists. However, this was not a show for the tourists as much as it was *for Ali's peers* – other Göreme hoteliers and restauranteurs. Ali had previously told me of his awareness that, because of his evident business success, others would likely 'copy' his business practices. As discussed in Chapter 2, the tendency in peasant-economies to 'compete to remain equal' by copying others' apparently successful innovations is strong in Göreme's tourism. Being aware of this tendency, Ali started, during the 2000s, to deliberately create certain effects in the relationship between tourism and culture. In other words, he knew that, because he was a role model of business/hotelier success, if he put *pekmez* on the breakfast buffet, bought gardens to grow produce for his hotels' restaurants, and renovated pigeon houses so as to have organic fertiliser, then others would likely follow suit. Indeed, many did and so, to some degree at least, Ali's plan worked.

Ali has had many such plans during the past two decades. In 2005, he told me that because there were so few animals such as donkeys and horses being kept by villagers, he had suggested to the mayor of that time to give enough coal to last through the winter to anyone who continued to keep such an animal; "That would keep the practice of using donkeys and horses to go to the gardens alive", he said. At that time he also restored some of the communal stone ovens in the old mahalles so that women would continue to use them to bake bread and roast pumpkin seeds. In addition, he and other hoteliers joined together with some foreigner residents to undertake beautification projects in the old mahalles, such as covering concrete-block walls with local stone and restoring communal water taps, troughs, and an old mill. Their expressed intent was to retain and present aspects of the old mahalles which tourists would find interesting and aesthetically pleasing. They saw such 'local' cultural elements as valuable for boosting tourism, and conversely, that if such elements were lost from the mahalles, then Göreme may no longer be attractive to tourists.

As already suggested, however, Ali's intent seemed not so much about using culture to boost tourism, but rather to harness tourism to boost 'culture'. Again, it is difficult here to decipher what stickinesses might be involved in such moves to retain certain cultural elements deemed worthy of preserving, although pertinent to this is Rowland's (2002) suggestion that any form of heritagisation 'implies a threat of loss and the need to preserve or conserve against an inevitable sense of deprivation' (p. 110). As is apparent from the discussion earlier in this chapter, sticky memories can contain the tug of nostalgia, but also the distress and shame of hardship and 'backwardness'. As it turned out, in 2009, the Göreme Mayor forbid residents from keeping cows in the old mahalles, a move some said he was right to do because of the shame of appearing backward in the face of tourism. Others said there were practical reasons for banning such animals, due to the smells and flies cows attracted potentially being a problem for nearby hotels. Nonetheless, there were many who did not want to lose these elements which had previously been integral to daily life in Göreme. For some, there lingered a sense of melancholy (*hüzün-lük*) at the perceived losses occurring around them. As Rosaldo (1989, p. 107) has suggested, it is often the case that nostalgic yearning is the 'mourning for what one has destroyed'. In the late 2000s, another Göremeli businessman – a long-time tour agency and car rental business owner – told me that everyone should move to the Yeni Mahallesi where people continued to be allowed to keep sheep and cows; "There they can have their culture back, away from the tourists and hotels, it will be better for everybody".

Unlike that agency owner, Ali, who was already becoming wealthy and successful in his business endeavours, clearly felt that he himself was in a position to be proactive in attempting to hold back, or perhaps even reverse, the changes occurring. When I chatted with him in 2009, Ali talked avidly of wanting to support and continue local food practices, such as dried foods, jams, and other preserves. He told me that the newly appointed Göreme Mayor was thinking about opening a kind of museum somewhere nearby Göreme in which he wanted to display 'heritage' items, such as costumes from the 14th and 15th centuries. Ali had gone to the Mayor to argue against this idea: "Why would you present the 14th or 15th Centuries? Let's try to keep what we have *now* – otherwise we're going to lose it!" Another related idea of Ali's at that time was to buy and renovate old houses and let local families live in them to run 'homestay' accommodation for tourists. As well as this bringing families back into – and thus reviving – the old neighbourhoods, he said that he envisaged tourists going to the gardens with their host-family and helping them; "Some tourists would pay to do that", he said. While he did not carry out this 'homestay' plan, the valley breakfast and the *pekmez* tour which Ali started in the 2010s were likely the manifestation of the parts of this plan which did eventuate. Along with his restoration of two hundred pigeon houses, the breakfast in the valley, the *pekmez* tour and the use of 'local' foodstuffs in hotels and restaurants were all attempts 'to keep what we have now', to ward off the threat of loss. More recently, the website of Ali's signature hotel reads:

> We always highlight the local culture and traditions. Our hotel guests get a chance to join activities like grape harvesting, cooking classes etc. Partnered with our travel agency, we provide all these cultural activities in our valley where we have an organic and traditional Anatolian farm.

Many hoteliers and restaurant owners – likely following Ali's example – have during the past 10–15 years increasingly promoted local food production by buying vegetables and fruit from local families, as well as by using locally made preserves and other specialities on breakfast buffets and restaurant menus. Hotelier and restaurant owners would frequently ask their mother and other female relatives to make preserves and other local food items for their business, while others paid women in their neighbourhood, or retired people such as Abbas and Senem, to make the foodstuffs for them. Additionally, other 'organic breakfast' restaurants appeared in the valleys adjacent to Göreme and those, too, showcased local specialities and used locally grown organic produce. Such use of local food products in tourist restaurants and hotels is broadly regarded in the tourism literature as good 'sustainable' tourism practice, not only due to its propensity to enhance tourist experience of aestheticised "localness" (Hall, Mitchell and Sharples, 2003), but also in terms of its supporting of local food systems while also helping distribute tourism earnings more widely across destination communities (see, for example, Everett, 2016; Sims, 2009). Indeed, the purchasing of *pekmez* and other preserves from Göremeli neighbours is often done explicitly in order to provide a small income to those still willing and able to undertake such food production activities, and also in acknowledgement that doing so will help maintain such food production knowledge and practices regarded as traditional to the area. Another welcomed consequence was the maintenance of a 'fallback' position for when tourism business would experience periodic down-turns, such as during periods of political instability in Turkey and, most recently, the Covid-19 pandemic. The continuation of such food production activities for tourism could thereby be said to act towards sustaining a locally embedded food security system.

Concurrently, however, in Göreme the involvement of such food products and activities in tourism engendered a revaluing which acted to legitimise the continuation of certain practices largely undertaken by women. In other words, these tourism-related processes of aestheticised, heritagised nostalgia could be said to act to shape women's lives by tying their daily activities "to a logic of sameness" (Gedalof, 2003, p. 92). It is pertinent to recall here Abbas's niece Sonbahar pointing out that her being sent to high school before it was considered normal for girls to do so had given her the lucky break of *not* having a life of tending to the fields and making *pekmez*. Similarly pertinent were the other women who told me that whereas girls used only to learn how to work in the fields, now *they want to study*. Just as Roy (2004) has argued that the aestheticisation involved in the process of revalorising and revitalising certain aspects of "localness" can serve to hide, or even render invisible, "the brutal mechanics of capitalist valuation" (Roy, 2004, p. 65), so urging the continuation of those activities identified as "traditional" and "local" in Göreme may directly undermine, in particular women's, hopes of becoming other.

Hence, as well as being overtly generational, sticky memories, hopes, and dreams are significantly gendered. They also have a way of working constantly to undermine and destabilise each other. During a 2010 fieldwork visit, a family I encountered in Aydınlı Mahallesi during their *pekmez*-making included a young woman who told me that she was studying to become a nurse at Kayseri University. She told me that earlier in the day, a TV film crew from Istanbul had come by and filmed them. Such scenes would likely ignite nostalgic interest among urbane Turks from other parts of the country. She went on to tell me: "The semester starts on Monday and I would like to have gone back early except that I still have work to do here in the village". Listening in, her mother responded by telling her this problem would not occur in the future since she would stop making *pekmez* because it was too hard. Another elderly woman in the group then joined in the conversation by saying that she loves making *pekmez* "because it is natural (*doğal*) work". This conversation clearly conveyed the felt ambivalence among women concerning *pekmez*-making practices.

Ali, too, displayed a sense of ambivalence by often becoming melancholic while reflecting on what he saw as the demise of Göreme culture. Along with his awareness, and desire, that the use in tourism of local food and local food practices, such as *pekmez*-making, would serve to revalue and revitalise these foodstuffs and practices and thereby likely urge their continuation, Ali also acknowledged the contradictions in *his own roles* in the tourism and cultural change relationship. During our talk in 2009, he had sounded somewhat apologetic for buying so many old houses which he incorporated into his hotel rather than converting them for villagers and homestay accommodation. He acknowledged that others complained about him making his hotel bigger: "They say I am just a capitalist!" He continued, "I also don't like capitalism, but you need to make money to do these things, to restore the broken old houses, and to support the culture, the farming and so on. So what can I do?"

Indeed, during my many talks with Ali over the past two decades, it was evident that he experienced considerable anxiety in relation to the "threat of loss and the need to preserve or conserve against an inevitable sense of deprivation" (Rowlands, 2002, p. 110). Just as Ahmed (2004, p. 124) discusses anxiety as being a feeling of uneasy anticipation of a threatening but vague event, it appears that such anxiety is the stickiness which 'moves' Ali to embark on his tourism-for-culture plans and activities. Moreover, since such a threat of loss has come to hang in the air in Göreme, as a spectre, it is interesting to consider what similar resonances there may be between the nostalgia proffered by Ali and the sticky memories of Abbas, and the fear of loss felt by Abbas's niece Leyla. While all of them contain ambivalences, it seems that, compared with Abbas's visceral, sticky, and attached nostalgia, the aestheticised nostalgia for Ali and Leyla is more detached; just as Leyla had left the *köy* (village) to travel and to become 'something other', so in some senses had Ali. Hence, if nostalgic yearning is the 'mourning for what one has destroyed' (Rosaldo, 1989, p. 107), then perhaps the threatening but vague event anticipated is the spectre of dreams come true.

The spectre of dreams come true

In the late summer of 2019, I went to help Abbas and Senem in their garden up on the ridge beyond Aydınlı Mahallesi. We went there in Abbas's car, driving up the winding street passing the old mahalle's many hotels, including Aydınlı Cave-hotel, the various hotels of Ali and, at the very top before heading out of the village along the dusty track, Osman's large new hotel. The place of the garden belonging to Abbas and Senem we went to was nearby the spot where the tractor and trailer carrying Ali's breakfast guests had stopped for Ali to begin to tell his stories about times gone by – when everybody lived from their gardens. Setting about harvesting Abbas and Senem's garden produce of tomatoes, apples, walnuts, and grapes, after an hour or so we took a break from our work and stood chatting while picking at bunches of the sweet black grapes and eating walnuts which Abbas cracked open between two rocks. Abbas looked out at the magnificent view over Göreme this garden afforded, and said:

> Before tourism, everybody in Göreme lived off these grapes; everybody had gardens and dried the grapes and then went to town to sell them at the market, and while at the market they would buy whatever they needed. These grapes are what everybody lived on… and now there are hardly any left.

He then looked further up the ridge and pointed out all of the gardens now owned by Ali, some on top of the ridge and some in the ravine below. Having the gardens pointed out in that way prompted Senem to exclaim: "Zengin!", meaning "He's rich!"

Wanting to thank Ali for inviting me along to the breakfast and *pekmez* tour, a few days afterwards I visited him at one of his hotels. As we were served tea, I told Ali how much I – and reportedly the other guests – had enjoyed our breakfast and *pekmez* experience. I then moved the conversation on to Ali's broader endeavours and asked if he felt optimistic about the future. He responded:

> I have no optimism at all. Yes, people are getting rich, we are making good money, but at the same time we are losing Göreme, losing our culture, losing how to relate with each other, how to care for each other.

He then added, "Even if everybody's making good money, we can never buy that back, *we can never buy Göreme back!*"

In the words of Ali and Abbas, there is poignant nostalgia, 'a sentiment of loss and displacement' (Boym, 2001, p. xiii) from them both, and yet their forms of nostalgia differ in important ways. Abbas's melancholic nostalgia is at once based on and conjuring of sticky memories – memories of times gone by – of a life now lost. The nostalgia of Ali, on the other hand, is more future-oriented and, with a sense of urgency about it, aligns with a kind of nostalgia which Bradbury has described as being 'not so much a longing for the way things were, as a longing for futures… that have been foreclosed by the unfolding of events' (Bradbury, 2012, p. 341). The nostalgia of Ali, then, might be seen as a longing for a future Göreme already foreclosed by dreams come true. A further possible form of nostalgia pointed out by Bradbury (2012,

p. 341) is 'the desire not to be who we once were, but to be... our potential future selves'. This nostalgia is future-oriented, too, but rather than being about a longing for a future which can no longer exist, this form is more related to hope: hope of being something *other than a pekmezci* (*pekmez*-maker), such as that conveyed by Neriman's daughter Zubeyde. It is little wonder that Ali's attempts to recreate stickiness by harnessing new regimes of value available through tourism have merely manifested as further contradiction and ambivalence with regard to sticky memories, hopes, and dreams.

Indeed, there are many people for whom dreams have not (yet) come true; for whom Göreme represents, at best, a hopeful space of 'the-not-yet-become' (Everingham, 2016, p. 522). They include those who come from outside of Göreme to seek work, such as the large number of women from nearby towns who arrive by bus every morning to clean and cook in Göreme's hotels, or the refugees – mainly from Afghanistan – who, during the past decade or so, have wound up in Göreme because of its plentiful tourism work. An example was Zaafirah, whom I met during the summer of 2019 when I popped into a café on Göreme's mainstreet to have lunch. Assuming from appearances that she was Turkish, I spoke Turkish to ask the waitress what lunch options were available. She politely replied in English and told me that she did not speak any Turkish. Since there were not many customers, I had a chance to chat with Zaafirah when she brought me my lunch. She told me that although originally from Afghanistan, she had spent some years studying English language and translation at a university in Tehran. She would have liked to stay in Iran to work but was unable to extend her visa there. She decided against returning to Afghanistan and instead to seek a better life in Europe. I asked how she had ended up in Göreme and, misinterpreting the meaning of my question, she said that she had walked – over the mountains from Iran and into Turkey; "It was VERY hard!", she kept repeating, "...you can't imagine!" "We walked in snow this deep", she said, holding her hand at hip level. Upon hearing through contacts that there were jobs for English speakers in Göreme's tourism, Zaafirah had arrived there, and found her first job as a housekeeper in a boutique hotel. She left that job because the manager had frequently shouted at her for, supposedly, not doing her job properly: "She just kept shouting at me", Zaafirah told me: "Why didn't she tell me in a normal voice? I am human, and, yes, I'll work for you, but don't treat me like this; I'm not your slave", she said. Things were better at her café job, although she worked twelve-hour days seven days per week, with the only exception being able to start work a little later on one morning per week because a requirement of her migrant identity card and ability to legally work in Turkey was going weekly to the police station in Nevşehir to having her fingerprints taken. Two months after I first met her, Zaafirah told me she would be leaving Cappadocia very soon because she had found people who would help her cross the border into Europe; she had an arrangement to meet them at an Istanbul location at midnight on a particular date the following week. I gave Zaafirah my contact details, but I never heard from her again; I thus will likely never know whether her dream of reaching Europe was fulfilled.

The story of Zaafirah's episode in Göreme is the story of crossing paths and juxtaposing hopes and dreams. For her, Göreme was a stepping-stone in her endeavours to reach Europe. As a stepping-stone, Göreme might be considered a place of hope in her striving towards a future-not-yet-determined, although I could not discern much hope during my interactions with Zaafirah. Rather she exuded a sense of precarity, of having been pushed out to the edge of things, to 'the nervous edge of an impossible paradise' (Little, 2020); it was this precarity that was being capitalised on in order that others in Göreme could build and make manifest their dreams. Afghanis had come to be deemed 'good workers' and, unlike Syrian refugees, were sanctioned by the Göreme Belediye office to work in the town. With fewer rights, however, their cheap labour could be easily exploited by Göreme's business owners. In 2018–19, there were many such workers, mainly men, in Göreme's tourism – doing the work that the young men and boys of Göreme used to do in the earlier days of tourism, when Göreme was still largely a hopeful space of the-not-yet-become, before dreams were realised.

That dreams have come true – even if only for some – raises a spectre because, relative to hopes, dreams may be less prone to being blocked by stickiness; unweighted and unanchored, it seems, dreams can go anywhere – they are without limits. To recall Ahmed's words shared at the start of this chapter: 'When a sign or object becomes sticky it can function to 'block' the movement (of other things or signs)' (2014, p. 91). The spectre of dreams come true is thus the anticipation that the stickiness of sticky memories might become entirely unstuck; as Abbas lamented, "how can the children learn if they only stay for half an hour?" While Abbas's generation had a higher propensity for such sentiments, the desire for sticky memories not to become entirely unstuck was also at times expressed by younger Göremeli. For example, when chatting one day with Osman's son, Numan, in his father's 'dream' hotel, he told me:

> We were brought up by our grandparents – we knew their struggle. They had a hard life, so we need to not throw them off our shoulders. We need to appreciate the difficult life they lived for us to have this now.

A similar sentiment came from Refik, who told me:

> I once sold a garden to pay my (café's) rent, but no more. Everyone is trying to look forward, but I want to look behind, back. I want to continue my father's traditions in his honour, I want to make *pekmez*, because nobody is doing that anymore.

Once again *pekmez* is conjured to encapsulate the continuing of past traditions, and also, importantly, a sense of honour and integrity. The *pekmez* stories in this chapter have acted as a poignant way to tune into the affective entanglements and ambivalences of change in Göreme. Without doubt, *pekmez* is ambivalent in its stickiness. Its stickiness can function to bind things, and people, together; those who have left Göreme and moved to Nevşehir, or Kayseri, or even as far and wide as Germany, Holland, or India, like to return at *pekmez* time, when *pekmez* fires once again become focal points for neighbours to gather around, and when the mahalle feels like a neighbourhood once more. Relatedly, however, *pekmez*'s stickiness may also act as an anchor, weighing things down and stopping them from moving, such as young women's hopes of becoming other. For if hopes are a striving for a future-not-yet-become, some layers of that striving may be resisted or undermined by sticky memories and the moral imperatives associated with them. On the other hand, sticky memories can create a 'desire not to be who we once were' (Bradbury, 2012, p. 341) and thus prompt hopes of *something better*, to become something other. Sticky memories, hence, can both propel *and* disrupt hopes of becoming other. Either way, hopes bear some stability in that they tend to be anchored to various known contingencies.

In contrast, dreams might be considered altogether unstable, and hence foreboding, precisely because they are relatively free from limits. In this sense, the spectre of dreams come true is exactly concordant with the spectre of unlimited change. Encapsulated in Ali's saying "Yes, people are getting rich,… but at the same time we are losing Göreme, losing our culture, losing how to relate with each other, how to care for each other", is not only a sense of nostalgia but also what Albrecht (2005) has named 'solastalgia'. While akin to nostalgia, solastalgia is much more about *place-based* distress and the sense of lost home even when still at home. Solastalgia is argued by Askland and Bunn (2018) to be the experience of a powerful temporal rupture wherein the sense of connectivity and continuity of "home" is broken. This rupture can be seen clearly in Ali's exclaiming further that: "Even if everybody's making good money… *we can never buy Göreme back!*" Such is the spectre of dreams come true.

Notes

1 Similarly, for Dawney (2011), 'affect offers a means of geographical analysis of *what is at work*: what resonates through bodies as a result of their historical imbrications of material relations, and of what these resonations can *tell us* about those relations' (p. 599).
2 Or, as Öztürk *et al.* (2018) would say, his becoming part of the 'new peasantry', hence, Öztürk *et al*'s suggesting that the 'new peasantry' necessarily both resists neoliberal globalization and exists because of it.
3 Soares' referencing of Bhabha's (1984) mimicry in colonial encounters – whereby the colonized perform their own fantasies about the colonizer in order to overcome their sense of inferiority – brings to light some parallels between colonial and tourism encounters.

8 "Enough! (*Yeter!*)"

Stories related to the spectre of dreams come true suggest a message – a warning perhaps – to be careful what you wish for. This message appeared as the moral-of-the-story in the short film *Jonah*, introduced in Chapters 2 and 3: when the dreams of two young men come true and their capturing a photo of a huge fish leaping out of the sea leads to rapid tourism development as well as to their own fame, the small African town where they live soon becomes a sleazy hellscape, bereft of any sense of the place it used to be, while the men themselves are left forgotten and forlorn. Mehmet – who after watching *Jonah* commented that Göreme's story was the same, except "instead of the fish, it's the balloons" – once carried his own tourism fame; in his Fred Flintstone characterisation, he regularly and proudly proclaimed that he was "born in a cave", and he named his – popular throughout the 1990s – bar 'Flintstones Cave-Bar'. Now, like *Jonah*'s lead character, Mehmet, too, at times seems forlorn, and tired; the hours of restful sleep are not long enough these days, sandwiched as they are between the music emanating from nearby late-opening bars and the rushes of gas released into balloons as they pass by bedroom windows in the early mornings. Mehmet's likening of Göreme to *Jonah* is thus a reminder that, while the story of Göreme's localised dynamics of global tourism, neoliberal capitalism, and other global fluxes is *Göreme's story only*, this story resonates with a great many places around the world. These resonances are precisely why it is so important to listen to, and to tell, the stories which my long-term involvement in Göreme enable me to tell. Equally important is to attend to these stories – melancholic stories, hope-filled stories, stories of how hard life used to be, and yet, equally, how hard life is becoming – without recourse to generalisation or attempted 'weighing up' of gains and losses, costs and benefits. Attending to these stories is perhaps the only way to really *know* the experience of living with tourism, and living with change more broadly, in the contemporary neoliberal world.

Mehmet has always been good for a story or two, be they stories from his childhood initially spent in Göreme and later growing up in Holland, stories from his years of driving large haulage trucks from northern Europe through to Iran, or stories of his tourism days playing at being "Fred Flintstone" for

DOI: 10.4324/9781003011200-8

the backpacker tourists who drank in his bar and stayed in the pansiyon converted from his family's old cave-house. Mehmet's stories tend always to convey his passionate attachment to his Göreme roots, whilst also suggesting a strong sense of worldliness, derived from his many and varied routes. I have noticed in Mehmet's stories, too, his frequent uttering of the word "enough"; either expressing that he *has* enough, sufficient, and does not need more, or exclaiming that he *has had* enough, usually referring to Göreme's reaching a level of things which is formidable, perhaps even unbearable. In the latter case – when exclaiming "Enough!/*Yeter!*" – he, like others, has a pleading tone in his voice, as if pleading for the march of things to come to a halt, whilst knowing of course that it will not. By entitling this final chapter "Enough!/*Yeter!*", I want here to draw together and develop key themes arising from the stories told throughout the chapters of this book, while also drawing some threads through from *Living with Tourism (1)*. In doing so, I depict the major strands, or stories, from all four decades of Göreme's tourism and social change as culminating in three crises: a crisis of enough, a crisis of hospitality, and a crisis of "our town". I will begin by discussing the emerging 'crisis of enough', which I see as being at the heart of the image of unlimited good developed in Göreme during the past two decades, as well as central within the looming spectre of unlimited change.

A crisis of enough

We got talking about Mehmet's childhood when he found and showed me an old photo of himself standing with other small children: "Look at my trousers, we were poor then". Mehmet was born in 1960, in a cave-room in the fairy chimney which is now part of the pansiyon that was created out of the old family house. When he was a small child, his father went to work in a mine in Germany, and then moved to Holland to work in a factory. His father was therefore largely absent for the first ten years or so of Mehmet's childhood, until, in 1971, his mother and the children went to live in Holland too. At first, there was confusion over which school year Mehmet should be placed in because his birth papers showed him as 7, rather than 11; as with many families in Anatolia, his parents had registered his birth four years late so that he would be slightly older and wiser when he became the age to do his compulsory military service. After completing high school and a motor mechanics apprenticeship in Holland, Mehmet became a truck driver, transporting flowers throughout Europe and sometimes through Turkey into Iran. The family had kept their old cave-house property – situated at the start of the road to the Göreme Open-Air Museum – and when backpacker tourism was developing in the mid-1980s, Mehmet returned and converted the old house into a simple pansiyon. The pansiyon opened for business in October 1986 and for the next few years, Mehmet returned to Göreme during the summer months to open and run the pansiyon. In the autumn, he would close the pansiyon and return to his truck-driving job in Europe for the winter. Having dual (Turkish and Dutch) citizenship, he could do this dual life easily.

I first met Mehmet in the mid-1990s, by which time, he was running his pansiyon full time. Speaking four languages fluently and employing considerable irony whenever he introduced himself to tourists as Fred Flintstone, he came across as a cosmopolitan, 'worldly' man, highly skilled at playing upon, or playing *with*, how tourists saw him and others in Göreme.

Upon Mehmet's father's retirement, his parents returned to Turkey full time and built a large house for the family on the outskirts of Nevşehir. This was around the time when Mehmet's children were of an age to go to high school, and living in Nevşehir allowed them to easily attend a private college there. For Mehmet, living between the Nevşehir house and the pansiyon in Göreme allowed him to have the best of both worlds. He further built up his business portfolio through developing shop buildings on Göreme's main street, plus starting a tour agency business offering bespoke, tailor-made tours around Turkey, usually guiding the wealthy international clients by himself. During the 2000s, he embarked on other business ventures elsewhere in Turkey, but after a few years came back, as he always had done, viewing Göreme firmly as his home. Over the past decade or so, Mehmet has rented out the original pansiyon but has continued to provide and guide tailor-made tours for international visitors to Turkey, whom he quickly comes to refer to as his "friends". With the pansiyon rented out, he decided to build a small house above his rented shops so as to have somewhere to stay in Göreme. Additionally, he bought and developed a cave-hotel in a nearby, much less touristic, village, which he gave to his now grown-up son to manage.

In 2019, with property prices in Göreme having risen exponentially in recent years, Mehmet was offered, by an out-of-town businessman, US$4 million for his Göreme properties – the small house, the shop buildings, and the pansiyon behind. He said that he would never sell, however, since his accumulated property and wealth is for his children and grandchildren: "Everything is about them now, and besides, I could never sell the pansiyon because it is my old family house" – in front of which the childhood photo had been taken all those years ago, when "we were poor". In any case, he told me, the rental income he receives yearly for the pansiyon and shops is enough for his family to live comfortably: "Now I have enough", he said, "and, in any case, what would I do with US$4 million"?

In Chapter 2, I established a core theme of this book as being about what occurs when peasant sociality, derived from an 'image of limited good', becomes entangled through tourism with the spectre of unlimited growth cultivated by neoliberal market economics. Mehmet's going from "Look at my trousers, we were poor then" to the rhetorical "what would I do with US$4 million?" points to a key aspect of this ongoing entanglement; that is, the notion of 'enough' becoming increasingly elusive. Of course, it could be said that Mehmet, like many Göremeli people these days, has more than enough. Others, those who have not accumulated such wealth and property, including contemporaries of Mehmet, tend to say nonetheless that they too have enough. For example, Refik who was mentioned in the last chapter in regard to his giving up his café on the main street because the rent had become too high, said that he did not understand why some people keep wanting more: "Why do they need a fancy car or a luxurious house, other than to show off?", he said. Then he continued, "I have enough. I have food to eat, a nice place to live, gardens to grow vegetables and grapes for pekmez; I have enough". On the other hand, in discussing the concept of 'enough' with the younger generation of Göremeli, those who were born around the time of my earlier fieldwork in the late 1990s and who now manage the tourism businesses of their parents, I was told that: "In tourism, you never say enough". This statement encapsulates well the relentless desire for *more* cultivated by neoliberal capitalism's notion of unlimited good.

Relative to this apparent increasing desire for more, for growth, the image of limited good orientation which had underpinned Göreme's early tourism days enabled 'enough' to be held much more clearly in view. To refer again to Foster's interpretation of peasant limited good beliefs, when it is understood that 'there is no way directly within peasant power to increase the available quantities [of good]' (Foster, 1965, p. 295), then life is lived simply in the pursuit of enough. Peasant sociality ensured that when one had more than enough, when a family's garden produce was especially abundant for example, along with preserving some produce to last throughout the winter, they would be sure to share their produce around; indeed, it may be that next year, another's garden would be more abundant than their own. Accordingly, within the first two decades of Göreme's tourism, Göreme peasant-entrepreneurs' hybrid business practices were heavily influenced by image of limited good beliefs. At that time, not only was the size of the metaphorical tourism pie viewed as being fixed and finite, but its limits were tangible and visible. Before the building of the Cappadocia airports and prior to internet booking platforms taking hold, almost all tourists coming to stay in Göreme arrived off buses in the small bus station in the centre of the village, and so their numbers were literally visible. As they stepped off the bus, they were invited into the bus station's 'Accommodation Office' to peruse the poster advertisements of all of Göreme's pansiyons and hotels – with all the posters being of equal A3 size to ensure a level playing field – in order to 'fairly' choose their Göreme accommodation. The Göreme entrepreneurs also formed price-fixing associations so that they would not continually undercut each other in order to attract more of the finite number of tourists to their own business. As it was described in Chapter 2, each year a minimum price for pansiyon rooms and day-tours was set, at the level of *enough*, so that the peasant-entrepreneurs were co-operating in their competing for an equal share of the tourist pie. When the size of the pie is visible and known, and as long as there is fair and equal sharing, also visible and known is the size of 'enough'.

In contrast, with the emergence of a notion of unlimited good comes a much more ambiguous and elusive sense of 'enough'. During the 2000s, and even more so during the 2010s, the growth of the hot-air ballooning sector and the rapid increase in social media influence and internet booking platforms meant not only an increase in the number of tourists visiting and staying in Göreme (the size of the pie expanding), but that the number of tourists – booking via the internet and coming directly from their arrival airport to their hotel – became less 'visible' (the size of the pie expanding, but unknown to what extent). Moreover, the 'tourist good' available started also to be viewed as expand*able* (the pie as "grow-able", without limits). Within this context, there has arisen a marked difference between the various generations of Göremeli tourism entrepreneurs. While the younger generation have more fully adopted the image of unlimited good and therefore "never say enough", for the older generation who can more firmly be considered hybrid peasants-entrepreneurs, the ingrained limited good beliefs and concomitant peasant sociality mean that 'enough' can still be held in view. This was illustrated to me by many people of that older generation, such as Mehmet and Refik quoted above, and also by Mustafa, the restaurant owner introduced in Chapter 2, who told me:

> I could have expanded my restaurant to include the cave next door, but I didn't want to. I have enough. Everybody else has got greedy, they always want more. But I don't want that. My restaurant and garden house are enough for me.

Noteworthy here is the relationship between the concepts of 'enough' and 'greed'. While levels of tolerance for conspicuous wealth have undoubtedly expanded during the past two decades, considerable tension has continued to exist in relation to conspicuous business growth. This tension is likely a hangover from the 'zero sum game' previously operating in Göreme's tourism which inhibited the motivation for growth precisely because, in line with Foster's notion of limited good, any one person's advantage was viewed as gained at the expense of all others. As tourism morphed and grew during the 2000s and 2010s, some businesses and individuals began to get markedly ahead of others, with not only the more successful business owners becoming wealthy, but the size of their businesses and business portfolios growing exponentially. Contrary to the level playing field which had characterised the earlier years of Göreme's tourism, significant inequalities emerged and, while a few villagers became multi-millionaires, those who did not have ownership of a business often struggled to make ends meet. Adding to the increasingly unequal landscape has been the growing number of non-Göremeli-owned businesses being developed, both in the central areas of Göreme and in the old neighbourhoods. These create considerable unease, sometimes propelling Göremeli business owners' desire to further their own business growth, and sometimes causing them to rally together in support of each other's business success, for example, by lending each other money to purchase a particular property in order to block an outsider business-owner from buying it. In this context, which is full of ambiguity and ambivalences, deployment of 'enough' may be a form of coping mechanism, especially when stated alongside accusations of others' greed.

Indeed, Mustafa's accusation of greed in the above quote, being directed at a generalised 'everybody else', had the likely purpose of bestowing relative honour upon himself. Moreover, if he had accused any particular persons of greed, the accusation would have directly grated away at those persons' honour, and hence clearly links to the peasant moral economy and associated 'competing to remain equal' principles discussed in Chapter 2. Throughout the four decades of Göreme's tourism, if anyone, or any business such as a hotel, was seen to be more successful than others, they would tend to have gossip and criticism levelled against them, with the gossip usually containing accusations of the success having been achieved through devious or dishonourable means. An example I was aware of in 2019 was local hoteliers accusing the non-Göremeli-owned and apparently successful hotels in the upper parts of Aydınlı neighbourhood of widening their pipes in order to draw on a greater volume of the limited water supply coming into Göreme. The 'devious' outsiders were being blamed for the extreme water shortages Göreme was experiencing that summer and the local hoteliers said the accused hotels should be made to narrow their pipes so that the water would be shared more evenly. The trouble was that, by virtue of their being outsider business owners, they were outside of the moral economy and so could not be harmed by accusations of illegitimate business practice, nor, for that matter, by accusations of greed. For Göremeli business owners, on the other hand, being viewed as engaging in straight, fair, and honourable business practice was considered of utmost importance.

As well as the tug between honour and greed pulling on notions of 'enough', jealousy is another affect mode which is likely, perhaps even more than greed, to tug on 'enough' and render the size of enough ambiguous and contingent. That *kızkançlık*, meaning jealousy or envy, is an oft-used and frequently heard word in Göreme is unsurprising given its strong links to both the image of limited good and the folk concept of the 'evil eye', also prevalent in Göreme. Indeed, rather than being simply understood as a vestige of somewhat unsophisticated rural belief systems, many scholars (for example, Dundes, 1992; Magliocco, 2004) agree that evil eye beliefs have strong links to envy. It is likely, therefore, that evil eye beliefs carry a strong relation to limited good beliefs in that both sets of belief, through rendering jealousy, act as social equalising mechanisms. In Göreme, as the difference between business success and non-success became ever-greater during the past two decades, so outward showings of jealousy increased, with people frequently accusing others of engaging in jealousy-induced gossip and, sometimes, business-sabotaging tactics. Indeed, the rise of jealousy/*kıskançlık* is a common complaint voiced about tourism's detrimental effects on villager relations and sociality, an example of which was Zubeyde, the middle of three generations of women in the same household introduced in Chapter 6, saying, "I don't like the small-mindedness of the people here, their jealousy and gossip". In relation both to business competition practices and business growth, the differences between the roles played by jealousy and greed are obscured. Also further obscured is the concept of 'enough' itself so that, while for some, invoking 'enough' might be used as a counterbalance, or a coping mechanism, to deal with feelings of jealousy, for others, jealousy may serve to evaporate any ideas of having 'enough'; "In tourism, you never say enough".

Being interested in knowing why those whom I had not heard saying 'enough' felt a need for more, I spoke to the older generation owner of an ever-enlarging hotel in Aydınlı Mahallesi, who said that when he accumulates money, he needs to spend it on another house to incorporate into his hotel or on another hotel entirely, so as to not have money sitting in the bank. In this sense, business success leading to further business growth is an inevitable, self-perpetuating cycle. I asked a similar question to a younger generation hotelier who had previously told me that while Göreme had rapidly become much more famous than it used to be due to *Instagram* and other social media, it was not yet as known as somewhere like Machu Picchu and so there was still a need to advertise more. It was in response to my asking why there was still a need to advertise that he replied: "In tourism, you never say enough". He went on to explain that the more people there are seeking to visit Göreme, the more the businesses would be able to select their ideal target market; for example, his upmarket hotel would aim for the luxury-seeking tourists. Beyond hoping that Göreme would become something other than a *köy* (village), he continued by saying that his dream was that one day Göreme would be known as an elite destination, like Monte Carlo; "I'd like Göreme to be that kind of level one day", he said.

Immediately after talking about Göreme becoming another Monte Carlo, however, the young man reflected on the problems the tourism growth was creating in Göreme. Traffic congestion "in this little town" is becoming the biggest problem, he said, while the growing number of balloons and ATVs is damaging the environment around Göreme: "Everywhere is looking like an airport, all ugly and dusty, it's disgusting!" He then added that limits would soon need to be imposed, restricting traffic into the old neighbourhoods, and controlling the ATVs and balloon vehicle activities in the valleys. There have been periodic attempts to ban ATV groups from entering certain valleys in order to protect villagers' orchards and vineyards from the dust the ATVs create, as well as to protect the tranquillity in the valleys, but these restrictions have tended not to be adhered to. Indeed, the hot-air ballooning sector-related limit of one hundred balloons in the first take-off each morning is one of the very few limits imposed upon tourism in Göreme which has been adhered to.

Yet, despite there having been a growing animosity towards the balloons for several years now, as I discussed in the '*Nuisance balloons*' sections in Chapter 3, the restrictions imposed have tended to be safety measures rather than being out of any environmental or social concerns associated with the balloons or related aspects of tourism. In relation to the accelerated growth, throughout the 2010s, people raised similar concerns about the sheer number of hotels, especially in the old neighbourhoods, and the impending need for restrictions. An example was Hatice, the daughter of hotelier Osman, who said

> I think tourism in Göreme is good at this level, but I don't want to think about the future because I don't want to see more hotels here. The capacity is full. So it is good right now but it needs to stop at this level.

In most respects, however, aside from the broader restrictions regarding heritage 'protection' under the auspices of the Göreme National Park and World Heritage listing, tourism appears to have become an unlimited good, in relation to which "you never say enough".

A crisis of hospitality

During fieldwork in 2009, I was chatting with Mustafa in his Director's office at the Göreme Tourism Development Cooperative when a friend of his, Derviş, who had recently returned to live in Göreme after living in Holland for 40 years, came in and joined our conversation. Upon hearing what my interest was, Derviş was keen to voice his opinion about tourism and change in Göreme, saying: "Tourism has changed the Göreme people; they've lost their hospitality". He went on to explain that this was because "they only want money now from tourism". While Mustafa agreed that only wanting money now was in part the reason for *misafirperverlik* (hospitableness) going down, in his view there was another reason:

> Now so many people come here for different reasons, so we don't know how to behave towards them. Before, fewer people came and everyone knew that they were a tourist, and so everyone would go up to them and say "*Hoşgeldiniz*" (Welcome) and shake their hand.

He continued,

> But now, if I see you Hazel, for example, I don't know if you are a straight-out tourist, or a researcher/book-writer, or are you living in Istanbul, so what do I say to you? We can't tell anymore who is a tourist and who is something else.

The idea that levels of hospitality have gone down in Göreme because of tourism is not new; indeed, in *Living with Tourism (1)* I wrote: 'The central place of hospitality in villagers' narratives is indicative of the fact that this issue is at something of a crisis point' (Tucker, 2003, p. 133). In a similar way to Derviş saying in the story above that, in Göreme, "they only want money now from tourism", I was told during fieldwork in the late 1990s that, compared with a more genuine hospitality shown towards tourists in the 1980s, "Now we just smile at tourists to get their money" (Abbas, quoted in *Living with Tourism (1)*, p. 132). This overtly reflective sense of an erosion of hospitality, I argued, was 'felt as a loss of an integral part of villagers' identity at a variety of levels' (Tucker, 2003, p. 132) in that 'Göreme people take pride in their 'hospitable culture', and the concepts of *misafirperverlik* (hospitality) and *misafir* (guest) are central to villagers' discourses regarding themselves, their lives and tourism' (ibid., p. 122). As an exchange of honour (Herzfeld, 1987; Selwyn, 2000), hospitality has been argued to be the supreme virtue in Turkish society (Delaney, 1991), and it 'is a severe condemnation of a person or village to be thought inhospitable' (Delaney, 1991, p. 194). It is unsurprising, then, that a perceived erosion of hospitality in Göreme should be considered a problem, or a crisis even.

One aspect of the increased scepticism voiced by Göremeli about their own ability to give hospitality to tourists highlights the ambiguous entanglements between hospitality and the 'crisis of enough' discussed above. That is, as jealousy and greed rise so, it seems, does the "focus on money". For instance, it was discussed in Chapter 3 how, as the hot-air ballooning sector developed during the 2000s and many accommodation businesses came to make a large portion of their income from the commission they received from selling ballooning tickets to their guests, the focus and level of hospitality guests received was compromised. Some establishments even went as far as offering free accommodation for the first night of stay, but with a hard sell on ballooning accompanying the night's stay. In those establishments, the accommodation price jumped up substantially on subsequent nights, with the idea to push tourists on after they had been sold a balloon flight, thus making room for the next (potential ballooning) customers. This practice inevitably altered guests' experiences in the accommodation establishments in that they moved away from the generous performances of 'Turkish hospitality' that had occurred in Göreme's *pansiyon* accommodations in previous decades (Tucker, 2003).

The 'paradox between generosity and the exploitation of the market place' (Heal, 1990, p. 1) which is arguably ever-present in tourism, and its associated term, the 'hospitality industry' appears to be acutely felt in Göreme these days, although experience of and ability to navigate the paradox has varied. As discussed in earlier chapters in reference to the women who invite tourists into their cave-home in the hope of selling them jewellery or knitted items, in contrast to men, these women have tended to remain relatively isolated in their tourism entrepreneurial activities, and hence have not necessarily been afforded the same skills of navigating different systems of meanings in their encounters with tourists, including the paradoxical meanings within their 'hospitable' encounters with tourists. Meanwhile, Göreme men working in tourism who have a built-up skill of manoeuvring through multiple systems of meanings are able to reflect on the nuances of how offering hospitality to tourists has changed over time. According to Numan, who manages his father Osman's hotel, the decrease in hospitality is due to changes in the way that tourism now operates in Göreme, as well as in the way that Göremeli people operate within tourism:

> Before, our parents' generation were more original. They didn't know how to do tourism so they just did everything naturally. They didn't know the rules of tourism so they created their own rules, according to what they knew. For example, if people arrive early before check-in time, our parents would say come and have tea or breakfast with me, whereas we tell them they can't check-in until 2pm, like in the rest of the world. Our parents didn't learn tourism from books, but we have.

In discussions of these issues with Numan's father, Osman, blame was placed more on changes that have occurred on the part of the tourists during the past two decades:

> The tourists are completely different now – there's no relationship. Tourism used to be about becoming friends. Yes, they'd pay money but we had good times together. Now all they do is pay the money, sleep, and that's it! Hospitality is gone now, because they don't want it, they don't want hospitality. It is gone because of them, not because of us.

The changes Osman was referring to here were discussed in Chapter 4, where it was described how, from the perspective of their Göremeli hosts, the tourists previously were considered 'good guests' in that they sought adventure and connection with the Göreme people. Not only did Göremeli 'hosts' such as Osman enjoy spending prolonged amounts of time chatting, drinking, and generally entertaining the mostly international tourists, but as I noted in *Living with Tourism (1)*, through their 'imposing of a host-guest relationship, villagers [were] able largely to negotiate relations of equality and respect' (Tucker, 2003, p. 122). Even after the time of the less formal backpacker pansiyons with their *ad hoc* host–guest relationships, many owners of the longer-established boutique hotels which developed from earlier pansiyons were able to establish friendships with their tourist guests over the years. Over time, however, a fundamental shift has occurred, neatly summarised by Dinhopl and Gretzel (2016, p. 135): "Rather than fetishizing the extraordinary at the tourist destination, tourists seek to capture the extraordinary within themselves".

My longitudinal ethnography in Göreme has enabled me to see how this shift is experienced by local 'hosts': in Osman's words, "hospitality is gone now, because they don't want it, they don't want hospitality".

A key change being referred to here is, as discussed in Chapter 4, the increased use of digital communication technologies, including social media, in tourism. During the 2000s, digital communication technologies meant that tourists became more connected with 'elsewhere' – spending time on Facebook, or communicating with family or work back home – and this lessened their propensity for local encounters and connections. During the following decade (2010–19), the rise and influence of social media coupled with the growing fame of the hot-air balloons led to Göreme's tourism becoming predominantly about providing backdrops for selfie-photos and 'likes'. The many complaints I heard about these 'new' tourists, especially in 2019, suggested that tourists were no longer considered 'good guests'; the new tourists' 'purpose' in Göreme was no longer conducive to a congenial host–guest relationship. Hoteliers also blamed the rise of online booking agencies, in that tourists' booking via these online agencies, plus their ability to write reviews of the hotel after their stay, had altered the hospitality dynamic substantially. On this, Osman said that the review system meant that "you have to give good service, and a perfect breakfast", in order to maintain a high rating. His son Numan went further, saying:

> The reviews have changed the way we do tourism. We used to do things because of our tradition, our hospitality, but now we do things only thinking about the review, for the ratings, not from our hearts. It's somehow not allowing us to do our real job, so it's changed us.

Numan's words juxtaposed "hospitality from our hearts" against the review and rating processes of contemporary neoliberal global tourism, which is "somehow not allowing us to do our real job".

As well as being increasingly incorporated into these globalised neoliberal tourism and digital technology processes, the tourism businesses overall, and in particular the hotels, have tended to increase in their levels of formality and structure during the past two decades. In turn, the owners of businesses have become less involved in the daily running of their business, and this has likely contributed to the felt drop in hospitality. Many long-time Göremeli hoteliers told me of being tired and no longer having any interest in interacting with tourists; they consequently have employed managers in order that they themselves can take a more hands-off approach in running their business. Moreover, towards the late 2010s, increasing numbers of hotels owned by Göremeli were rented out, often to non-Göremeli. Not only was renting out your hotel deemed "an easier life", but it also reaped a substantial-sized lump sum – usually 60% of what the owner would earn from running the business themselves – at the start of each year with which the owner could buy or develop a further business. For the period of the rental agreement, the owner would usually have no involvement in the hotel at all so that the lease-holders and employees of the businesses – many of whom are from outside of Göreme – would do much of the 'hosting' of tourists. Göremeli people have thus gone from actively 'hosting' to 'owning' in tourism.

I wrote in *Living with Tourism (1)* that, during the 1980s and 1990s, there was 'a clear sense among Göremeli people that in their village they were "hosts" to their tourist "guests"' (Tucker, 2003, p. 118); as Mustafa said in the encounter above, "everyone would go up to them (tourists) and say '*Hoşgeldiniz*' (Welcome)". The roles of 'host' and 'guest' were overtly used by Göreme villagers in order to negotiate and determine their relationship with tourists: 'It is precisely the positioning of hosts and guests that enables the Göreme villagers to have a significant say in determining their interactions with tourists' (ibid., p. 125). At that time, strains regarding hospitality and hospitableness arose directly through villagers' dealings with tourists, and were therefore tensions *within* this 'host-guest' relationship, such as villagers' hospitality sometimes being abused by tourists, or it being rebuffed due to misunderstandings in the encounter. More recently, the crisis has shifted in that, as Mustafa indicated, "Now so many people come here for different reasons, we don't know how to behave towards them". Not only has it become increasingly unclear who is a 'guest' but, amidst increased overlapping mobilities, it has also become increasingly unclear as to who is a 'host'. The following encounter story illustrates these points.

One late afternoon in the summer of 2019, I was walking along the main street and came across Mehmet drinking a Turkish coffee at a table outside his friend's tour agency business. He invited me to join him and introduced me to his friend, the agency owner, explaining that she was from Urgup and was the daughter of an Urgup man I had known for many years since he had tourism businesses in Göreme in the 1990s. Upon Mehmet's request, an agency employee brought me a Turkish coffee. After a short time, two young men came along and were told that they needed to wait for ten minutes or so to be picked up for the ATV tour they had previously booked through the agency. Pulling up some chairs, Mehmet and I invited them to sit with us and we all got chatting. One of them, Sali, came from London and spoke with a thick cockney accent, while the other, Abdullah, came from Wales. Upon learning their names Mehmet asked, in Turkish, if they were Turkish, to which they replied that, while they had grown up in the United Kingdom, both of their families came originally from Kayseri (the large city about one hour's drive from Göreme).

We continued our conversation half in English and half in Turkish, and they complimented me on my Turkish language ability. Mehmet and I explained about our long-time friendship and my long-term association with Göreme. I happened to be carrying a copy of *Living with Tourism (1)* and Mehmet was delighted that he could show Sali and Abdullah a photograph in the book – of himself sitting in front of his 'Flintstones Cave-Bar' – from 20 years earlier. We next asked them about their trip and they told us that having just visited family in Kayseri, they were pleased now to be able to "escape" their families by travelling to Göreme and other parts of Turkey. After spending a couple of days in Göreme doing "adventure activities" such as ATV riding, they were planning to head to Izmir to try the nightlife there. After a while, an agency employee, also from Urgup and unknown to Mehmet or I, came out and explained to the young men in English that they would be picked up soon and transported to the ATV station in a valley from where their tour would begin. When Mehmet interjected to tell the employee that the two men were Turkish, he appeared very embarrassed and apologised profusely for speaking to them in English, to which Sali replied, in English: "It doesn't matter, I'm from London." We all laughed about the confusion and, coming out to learn what all the laughter was about, the agency owner responded, "Oh my God, you can't tell anymore who anybody is, or what language to speak". To this, Mehmet said that even though he was born in Göreme and had lived there for much of his life, when he walks along the main street these days, he is often asked, in English, "Can I help you?"; "It is true that in this town nobody knows anymore who anybody is!", he said.

Molz and Gibson (2007) have pointed out that despite the 'host-guest' metaphor being pervasive in tourism discourse, empirical studies consistently call the binary oppositions between the categories of host and guest into question (p. 7). Relatedly, Maitland (2021) notes that it is becoming 'increasingly implausible… to separate visitors and locals' (p.16), in part because: 'Conventional distinctions between leisure, business and visiting friends and relatives (VFR) are breaking down multi-purpose visits' (ibid. p.17). The English–Turks illustrate this point well. Moreover, Mehmet's being treated as if a tourist in his hometown demonstrates why, as Molz and Gibson (2007) suggest, it may be more useful to 'frame host and guest as fluid, contested social roles that people move into, out of, and in between as they negotiate extensive overlapping mobilities and social memberships' (p. 7). While this may be so, however, it is precisely these fluid and contested social roles that seem to have become so troubling in Göreme. As Bell (2007, p. 29) has pointed out: 'Hospitality – as a relationship marked by poles of host-ness and guest-ness, and by the obligations and rewards this bipolarity brings – is itself destabilized as we enter an increasingly mobile age, a society of mobilities'. Indeed, the growing sense of 'not knowing who anybody is anymore' would explain well why an erosion of clear 'poles of host-ness and guest-ness' might be experienced as a crisis of hospitality. As Mustafa, Director of the Tourism Development Cooperative, expressed, not knowing who tourists are anymore has rendered it difficult for Göremeli to be hospitable; being unsure of people's 'guest' status makes it difficult to know whether or how to welcome them. Moreover, while Mehmet struggled with his being considered, by non-Göremeli tourism workers, a tourist in his hometown, he and Derviş – both of whom have lived in Holland for substantial lengths of time – themselves call into question the binary of *Göremeli* and non-*Göremeli*, local and non-local.

Meanwhile, the great many tourism workers who these days come to Göreme from other Cappadocia towns or other parts of Turkey, or even as far away as Afghanistan, subvert this binary further in that, as well as their not recognising Mehmet as 'local', they may not be recognised, by tourists for example, as 'non-local'. As Garbutt (2006) has suggested, the idea of being 'local' is always contingent and in need of being actively performed. As tourism in Göreme has grown and morphed, then, the associated increased overlapping mobilities and 'contested social roles that people move into, out of, and in-between' (Molz and Gibson, 2007, p. 7) have undoubtedly called into question: 'Who gets to be a guest, and under what conditions? Who gets to be a host, and under what conditions?' (Molz and Gibson, 2007, p. 8). Molz and Gibson continue this line of questioning further by asking whether the host is necessarily a citizen of the host nation-state. In relation to Göreme, the pertinent question is whether the host is necessarily a citizen of the host town. As if in answer to this question, I was told by some Göremeli people that a key reason for levels of hospitality going down was: "All the workers coming in from Nevşehir and other places, they can't do hospitality, even the farmers in this town know how to do hospitality better". Yet, throughout the 2010s, every morning several bus-loads of workers arrived, mostly from the nearby towns of Nevşehir and Avanos, to work in Göreme. While some of their jobs – such as housekeepers, cleaners and cooks – were conducted largely 'backstage', other roles, ranging from reception and waiting staff to balloon piloting, did involve interacting with tourists in a 'hosting' capacity.

While many of these personnel would be assumed by tourists to be from Göreme, more likely is that the majority of tourist guests would not think about the 'citizenship' or 'local' status of those hosting them. Indeed, it may be that the neoliberal logic of 'hosting' tourists circumvents any such thinking. Yet, the presence of these hosts-from-elsewhere nevertheless render ever-present the question of who gets to be a host, and under what conditions. As Lynch (2017) contends, hospitable convention would have it that a sense of 'anchorage', or 'being-at-home', is a necessary 'pre-condition to allow the individual to act as host i.e. offering welcome to Others' (p. 180). However, just as being a 'local' is always contingent (Garbutt, 2006), so might a sense of 'anchorage' and 'being-at-home' always be relative, contingent, and unstable. For example, the sweetcorn seller featured in an anecdote in Chapter 4, who set up a stall on the main street during the summer of 2019, was from the east of Turkey and while he may not have had a sense of 'being-at-home' in relation to people he knew to be *Göremeli*, he might very well have felt a sense of 'being-at-home' when interacting with international tourists. This sense may then have afforded him, even if periodically and often only momentarily, a hosting subject-position, although it is precisely the momentary nature of his 'hosting' role which exposes 'the fragility or arbitrariness of hospitable conventions' (Rosello, 2001, p. viii; see also Bell, 2007 on 'Moments of Hospitality').

Another group – whose own precarity exposes the fragility and arbitrariness of hospitable conventions even more overtly – are the *"yabancı"* (foreigners) seeking temporary, or otherwise, refuge in Turkey who find work in Göreme's tourism. During the past decade, and with a substantial upsurge in 2015, Turkey has taken in and granted 'temporary protected' status to more refugees than any other country in the world. Among these, while Syrians have not been particularly welcomed in Göreme, many Afghanis and Iraqis have obtained jobs in hotels and restaurants in Göreme, such as Zaafirah who featured in the last chapter. I was unable to determine the mix of altruism versus utilitarianism – or even exploitation – in their employers' motivations for giving these *yabancı* (foreigners) work; much is to be said for the simple and ready availability of these mainly young men meeting a need for workers. Undoubtedly, they were cheap to employ and, with very little by way of workers' rights relative to Turkish employees, they could quite easily be exploited by their employer (again speaking to levels of 'hospitality' in Göreme). Nonetheless, some businesses retained the same one or two Afghani employees for several years, with the Göremeli business owners telling me they were pleased to be able to train and thus enable these young men to potentially open their own tourism business one day. Yet, all of these employees – even those who had been in Göreme for several years – raise poignant issues regarding the conditions of 'hosting'. When I encountered these employees, or was made aware of them by the owners of businesses, I usually found them to be withdrawn. This included Zaafirah, with whom I had several conversations during her two months of working in the café, who was always rather taciturn – listless even – as she went about her café work. The situation of Zaafirah and others from Afghanistan speaks to Rosello's (2001) discussion of how 'the parallels between the immigrant and the guest' (p. 8) place firmly into doubt the ability of the immigrant to truly ever perform 'hosting'.

Going beyond Molz and Gibson's (2007, p. 8) question of whether the host is necessarily a citizen of the host nation-state, Rosello (2001) points out, drawing on Derrida, that offering hospitality without the necessary precondition of 'anchorage' might be to *imply* a being-at-home: 'To dare say welcome is perhaps to insinuate that one is at home here... thus appropriating a space for oneself, a space to welcome the other,' (Derrida, 1999, p. 15, cited in Rosello, 2001, p. 17). Hence, the troubled experience for Mehmet when offered welcome by out-of-town employees of tourist restaurants on Göreme's main street; the employees were at once insinuating that they were 'at home here' while also, within this dynamic, insinuating that Mehmet was '*not* at home here'. This was undoubtedly quite a slur, especially given that, as Delaney (1991, p. 202) has said in relation to Turkish village-belonging, 'more than merely a place one lives, a village is an indelible part of one's being and identity'. Indeed, as referenced earlier, in Turkey, people 'belong to their village in a way they belong to no other social group' (Stirling, 1965, p. 29). This holds true for Göreme despite its having outgrown village size and population. Hence, in addition to the uncertain paradox faced by any 'non-citizens' (of nation or of town) placed in the position of 'host' to tourist 'guests' in Göreme, there is a significant conundrum for Göremeli business owners in their employing 'non-citizens' to host. According to the logic of hospitality, there is an impossibility of hospitality, or conversely, only a possibility of *in*hospitality on the part of 'non-citizens'; 'the ultimate logic of ethical hospitality means, in certain cases, that inhospitality is the most perfect form of hospitality' (Rosello, 2001, p. 12).

Yet, this logic assumes a clear demarcation between 'host' and 'guest', citizen and non-citizen, insiders and outsiders; a demarcation which is increasingly denied by the new mobilities arising from Göreme's becoming a significant place of global tourism. The increased movement in and out is perhaps compelling a necessary acceptance of ambiguities around 'citizenship' and 'hosting' ability, as well as the possibility of 'non-citizens' becoming citizens over time. Indeed, while many long-staying employees-from-elsewhere struggled to appropriate a space for themselves within Göreme's tourism, some have managed to develop their own business through which they are able to create 'a space to welcome the other'. An example is Fatih, featured in Chapter 7, who came from Aksaray to work in others' shops but now has his own carpet shop business through which he has become capable of 'manoeuvring through systems of meanings' (Vertovec and Cohen, 2002, p. 13) in his 'hosting' interactions with tourists. Another example is Yunus, from Avanos, who eventually took rental of the pansiyon he had worked in for several decades and, while he did not own the business as such, the rental arrangement placed him in the position of 'host'.

Other accepted ambiguities include the many "tourist-brides" who have married into Göreme families and who join in the hosting of (other) tourists in the businesses they run with their Göremeli partner. Following the 'romantic relationships' between tourist women and local men I discussed in *Living with Tourism (1)*, during the past two decades many such new and continuing liaisons have occurred, resulting in marriages, divorces, dual-nationality children, and Göreme families having family members, and in-laws, living all around the world. Even those who are divorced from their Göreme partner are okayed to insinuate "I am at home here", especially if they have children who are Göremeli. Other foreigners, too, who have bought houses in the old neighbourhoods, including myself, are often referred to as "local", or "Göremeli" (of Göreme).

Indeed, living with tourism for now over four decades has significantly changed the dynamics around who can be considered 'of Göreme' and who can perform welcome there. Whilst the 'crisis of hospitality' I discussed in *Living with Tourism (1)* was experienced by Göremeli 'hosts' as their hospitality 'becoming eroded *through their dealings with tourists*' (Tucker, 2003, p. 132), during the past two decades their direct dealings with tourists have vastly reduced, and so the crisis in the 'perverse dynamics' of hospitality, as Rosello (2001) refers to them, has become something different. Firstly, while many moments of hospitality undoubtedly do still regularly occur, such as the coffee encounter story above and many of the anecdotes told in earlier chapters, a neoliberal logic of hosting (serving?) tourists has largely overridden the ability to perform hospitality 'from our hearts', thus manifesting in practices of hospitality which are perverse according to local hospitable conventions. Moreover, by employing hosts-from-elsewhere, further perverse dynamics are created due to ambiguity, and perhaps even impossibility, regarding the capacity of those without anchorage to play host to tourists. While further ambiguity surrounds the ability for incomers to appropriate a space for themselves to perform welcome, the ever-increasing overlapping mobilities 'that people move into, out of, and in between' (Molz and Gibson, 2007, p. 7) have, in David Bell's (2007) words, destabilised 'the host-guest relation, leaving the identities of host and guest fragile, uncertain, decentred' (p. 29). The emerging sense of 'not knowing who anybody is anymore' is troubling because, beyond being a crisis of hospitality, it is, to use Herzfeld's (1987) words, a crisis of 'control over the metaphorical 'home' at all levels' (p. 81). This crisis of control over the metaphorical home links to there being, also, a crisis of "our town".

A crisis of *"our town"*

In the summer of 2019, I visited friends – two brothers – in the hotel that they jointly owned and, while we were sitting chatting in the hotel's reception area, a mutual friend Mehmet – the successful businessman and key *Instagram* influencer for Göreme introduced in Chapter 2 – came in and joined us. Mehmet had come to the hotel because he wanted to talk to his hotelier colleagues about the 'Big Bus': a few days earlier, the Big Bus Tour company had set up an operation in Göreme, with two open-top 'Big Buses' to run a hop-on-hop-off service on a set route around the main tourist attractions of the region. The three Göremeli men were clearly upset by the company's appearance, saying that it could hurt the business of the great many Göremeli-owned tour agency businesses which offered guided day-tours around a similar route. What they seemed most upset about, however, was the Big Bus operation having suddenly appeared apparently without obtaining permission from any Göreme party and without anybody even knowing who they were. The three men were talking in Turkish so agitatedly that I found it difficult to follow what they were saying, so for my benefit, Mehmet said in English,

> It's not good that someone starts a business in Göreme and we don't know anything about it; maybe the Belediye (council office) didn't even know about it. This is *our town*! They shouldn't just arrive here from nowhere with their big busses and nobody knows about them!

The three men continued their animated chat in Turkish, saying they thought Big Bus was a London-based company and that perhaps it had partnered with the Nevşehir governor in order to obtain permission to operate in Göreme. They then went on to discuss what they might do about it: "We can't do anything", one of them said. Then another said, "We can if we all join together and do something". Hassan, one of the two brothers said, "The only thing we can do is tell all the Göreme hotels not to recommend Big Bus or sell their tickets".

Figure 8.1 BigBus in Cappadocia.

Big Bus Tours is an international corporation with operations in multiple cities around the world, including London, Paris, New York, and Hong Kong. Not only does the opening of a Big Bus Tours operation in Göreme raise concerns amongst Göremeli of the type borne out of lingering 'image of limited good' beliefs – that is, concerns about Big Bus stealing from Göreme's limited tourism pie, but the conversation above highlights that it was the sudden appearance of the company in *"our town"*, without warning and apparently without the mayoral office's permission, that was experienced as such an affront. There is no doubt that the increasing number of hotels and other businesses, in particular hot-air balloon companies, being bought or built by investors from outside of Göreme during the past two decades has led to the sense of *"our town"* becoming more and more tenuous. In 2019, there were 28 hot-air balloon companies operating in the area, with over half of these under direct ownership or lease arrangement of Dorak, an Istanbul-based company.

However, that the Göremeli businessmen's not knowing about Big Bus was the cause of such agitation suggests that there otherwise continues to be a high level of monitoring and ability to know what is going on in *"our town"*. Moreover, the three men's envisaging of a kind of pact being formed amongst hotel owners whereby they would agree not to recommend or sell Big Bus tickets speaks not only to the point that there continued to be high levels – reportedly 80–90% – of local ownership of tourism businesses, but it also indicates a continued sense of 'Göremeli togetherness' vis-à-vis tourism in *"our town"*. This tendency towards collective action is likely a hangover from the business associations, discussed in Chapter 2, which the Göremeli entrepreneurs set up in the 1990s to counteract their own competitive tendencies. These associations engendered strong levels of cooperative behaviour and, although they no longer exist, it appears that the sense of collective self-determination they engendered among entrepreneurs lingers on to some degree.

This sense of collective self-determination has long been a characteristic of Göreme's tourism developments. With high levels of Göremeli-led tourism initiatives and entrepreneurial activity as well as significant decision-making afforded to the municipality office (*Belediye*) and its elected Mayor, I concluded that during the 1980s–90s, Göreme's tourism could be very much described as 'community-based tourism' (see Tucker, 2016b). This aligned also with broader ideas around peasant forms of production being characterised both by cooperative behaviours and maintenance of autonomy and control, as discussed in Chapter 2. Stirling's (1965) point about the Turkish village community being a key unit of sociality is pertinent here. As I explained earlier, although Göreme is technically a *kasaba* (a small municipality or town), with a population always around 2,000 inhabitants, it is just on the cusp between being a village and a town. In *Living with Tourism (1)* I myself and many of the people quoted referred to Göreme as a village and, while it is more often referred to these days as *"our town"*, many aspects of village-belonging and identity continue. Indeed, I have often noted in my fieldwork diaries that whilst, these days, Göreme is usually referred to as a "town" by Göremeli when speaking in English, in Turkish, they continue to refer to Göreme as a "köy"; I have never heard Göreme being referred to by Göremeli as a "şehir" (town).

The sense of collective self-determination among Göremeli had been significantly disrupted early on in Göreme's tourism developments when the Göreme National Park was formed in 1986. Due to Göreme village being situated in the middle of the national park area, it was immediately appropriated under national interests in such a way as to privilege a particular view of the area not only as a place for tourist consumption but also as in need of 'protection' for the purposes of tourist consumption. Consequently, as discussed in Chapter 5, all cave-houses and fairy-chimneys, and especially those in the old mahalles, came under preservation restrictions so that any alterations or building work required permission both from the 'Protection Office' in Nevşehir and the Göreme Belediye planning office. Having to deal with the Nevşehir office caused considerable antagonism for Göreme residents and entrepreneurs seeking to do building and restoration work, as I explained in *Living with Tourism (1)*. For example, I quoted one woman telling me when I commented on how dark her cave-kitchen was:

> Yes, but we are forbidden to make new windows or shelves or anything in the rock. Before it wasn't forbidden, but when tourists came here it became forbidden.
>
> (quoted in Tucker, 2003, p. 165)

Similarly, a tourism entrepreneur said:

> Suppose you have a cave... and your toilet is falling down and you want to build a new one...All those officials ...say you cannot do it...And probably they are from Ankara or Nevşehir, and they are just sitting at their desk...It's becoming a big problem. They say: "This is the rule, it is forbidden...". But it may not match with the people's life.
>
> (quoted in Tucker, 2003, p. 164)

A very different perspective came from the authorities, for instance, I was told by the Ministry of Culture official from the then 'Protection Office' in Nevşehir:

> The local people are not cultured; they are villagers, they are uneducated...The Göreme people don't understand our preservation project because they are uncultured. We cannot teach them how to protect the place – they make things ugly.
>
> (quoted in Tucker, 2003, p. 159)

This official's view bore striking similarity with Delaney's (1991) observation that Turkish government policy in relation to villages at the time of her field-work appeared to be 'informed by a particular view of peasants, namely that they are conservative, uninterested in change, and unintelligent' (p. 220).

The antagonistic relationship between 'inside' and 'outside' of Göreme has further resonance with Delaney's (1991) suggestion that 'villagers believe that all bad things come from outside' (p. 220). Also relevant is Delaney's (1991) suggestion that 'the village' in Turkey is symbolically female, and hence, in contrast to the town or city which is *açık* (open – and by implication tainted, polluted)' a 'village, like a proper woman, should be *kapalı* (closed, covered)' (p. 198). For this reason, the 'openings and closings' of the village should be monitored and controlled. Hence, in the above anecdote concerning the Big Bus company, the realisation by the three men that it is becoming increasingly difficult to monitor the 'openings and closings' of Göreme strikes as more than being a crisis of control over home. Rather, the realisation that the 'village' has become *açık* (open) is a crisis of *'home'* itself; it is literally a crisis of *"our town"*.

Indeed, well before the arrival of Big Bus, as well as the increasing presence of private companies and individual investors from outside Göreme, such as the Istanbul-based company, which purchased several hotels and balloon companies, significant governance changes in the late 2010s afforded external authorities increased power in relation to Göreme. Firstly, a law regarding the municipality system in Turkey which came fully into force in 2018 (Ozenen, 2018) removed decision-making power from small municipalities and centralised the decision-making authority to the regional city, in this case, Nevşehir. A year later, another change which further diminished the decision-making powers of the Göreme *Belediye* (Municipality Office) was the establishing of the 'Cappadocia Area Directorate', a newly established governance body ostensibly responsible for the 'protection' of the region (Özgentürk, 2019). Additionally, later in 2019, the government decreed the disestablishment of Göreme's National Park status. While, as mentioned, the formation of the national park had seen an initial disruption to a sense of collective self-determination in Göreme, over those three decades or so, the ways in which Göremeli who wished to build or restore property for business development should navigate the related 'protection' authorities had become known and understood. However, when in 2019 the 'Cappadocia Area Directorate' became the single authority body overseeing planning, zoning, and preservation in the Cappadocia region, it was seen as removing all sense of self-determination. In particular, the new so-called 'protection' body would completely bypass the local municipality office, thus removing any say in local zoning and planning that the Göreme municipality previously had. I was told by one Göremeli businessman: "We used to know who we would go to for (building) permission. Now this new organisation makes all the decisions. We lost control of *our town*".

Related to these governance changes in 'protection' of the area, some reportedly antagonistic incidents occurred in 2019 whereby the new presidential-directorate, as it became known, ordered the tearing down of all building structures in the area which had been built without proper consents. This included some tourism businesses which, often starting as simple stalls constructed out of wood, had been developed by Göremeli people over the years to become established businesses such as cafés and shops. I was told by several people about a tense clash which had occurred one day when it became known that bulldozers would tear down the cafés and souvenir-shop businesses situated along a viewpoint ridge overlooking Göreme. The businesses had been there since the 1980s–90s and the incensed villagers rallied together and assembled on the ridge to stand in front of the bulldozers to prevent this from happening. While that particular protest succeeded in stopping the demolitions, many older structures that housed cafés and other businesses were torn down.

Moreover, the new directorate was also intended, as a part of its 'aims to take measures for the protection of the famous region' (Alp, 2019), to 'cut the bureaucracy for those wanting to invest in the picturesque region' (Alp, 2019). A purpose of the new body was thus likely, in reality, about smoothing the way for outsider investors to move into Göreme's tourism business environment. In 2019, large hotels which were unsympathetic to local architecture and design conventions were appearing all over the region, including right in the centre of Göreme (see Figure 2.5). On this, Abbas talked to me about one large new hotel in particular, saying: "It's not good, it is ugly! It has broken Göreme because it is so ugly with its big black roof. It has completely spoilt Göreme and especially the view from Aydlınlı Mahallesi". He then said that I should write about it in my book; "We complain, but they don't listen, but maybe they'll listen if a foreigner says that it's ugly". In reference to the same issue, Ali, the hotelier who featured in Chapter 7 in relation to his own ideas about 'protecting' Göreme, complained:

> The authorities complain that the local people are fucking up Göreme, but one of them damages the place more than all of us together. We've been developing things slowly for 25 years, and then they come in and build ugly things, and because they easily get the license they can do it. Then they try to pull down these small wooden things which aren't nearly as bad as their big ugly hotels. We've lost control of Göreme.

Aligned with suspicions that the new commission's main purpose was to protect the business interests of external investors, there was speculation that the government's late-2019 decree to disestablish Göreme's status as a national park had a similar agenda. For over three decades, the environmental protections afforded by the national park status had prevented an influx of domestic and foreign investments into the immediate area in and around Göreme. The larger-sized hotels built by national and international hotel chains such as Dedeman and Hilton had been kept outside of the national park boundary, mainly in the towns of Nevşehir, Avanos and Ürgüp, all approximately

10 kilometres from Göreme. So, while the formation of the Göreme National Park had enacted something of an appropriation of Göreme's cave-houses and rock structures, and therefore, in part, the daily lives of villagers, paradoxically perhaps, the national park status had also maintained a form of 'protection' which had enabled Göreme's small-scale, incremental, and predominantly locally owned tourism developments. Regarding this issue, in a special boxed section entitled 'Saving Cappadocia' in the 2022 edition of the *Lonely Planet Turkey* guidebook, Jessica Lee wrote:

> Like many regions that have witnessed a tourism boom, Cappadocia walks a tightrope between hanging onto its authentic soul... and responding to the push for progress... However, in October 2019, in a move that horrified many locals, the government removed national park status from the region. Much of the Cappadocian community fears this could open up vast swaths of the area – previously protected as part of the Göreme National Park – to development. The next few years may prove critical in preserving Cappadocia's unique topography and cultural heritage.
>
> (Lee, 2022, p. 465)

Just as Göreme people's relationship with Göreme's national park status was always an ambivalent one, it is unsurprising that the dismantling of the national park status has been met with further consternation.

The growing crisis of "our town" is also, no doubt, related to increased displacement – both actual displacement and fears associated with 'place-based displacement' (Cocola-Gant, 2018) during the past two decades. In relation to actual displacement, the gradual loss of neighbours and familiarity particularly in Göreme's old neighbourhoods has manifested as an almost total transformation of these neighbourhoods into what some people have referred to as akin to a *'tatil köyü'*/holiday village. While, initially, only the central streets of Göreme were zoned for tourism business, during the 2010s, the old neighbourhoods were transformed from their previously being considered *kapalı* (enclosed) to being filled with businesses, tourists, and thus to all the openness-to-the-world that tourism entailed. Being increasingly unliveable for residents, these processes of transformation led ultimately to displacement of residents (neighbours) from these neighbourhoods. Those who could not afford the increasingly high land or rental prices in Göreme's Yeni Mahallesi were forced to move away to Nevşehir or Avanos.

It was in the late 2000s that Göreme's "housing problem" had started to become a major concern. In my 2009 fieldnote diary, I recorded a conversation with a tour agency owner, who had moved already to Nevşehir, about these concerns: "I had to leave my home town. There was no possibility to live here. Now I can't even vote in the Göreme (mayoral) elections, because I count as Nevşehir population now". Although this man came back to Göreme every day to look after his tour agency business, life lived in an apartment block in town was inevitably very different from how his family's life had been in Göreme. In contrast to men who might come back to Göreme daily to attend to their business, or simply to spend time in the Göreme teahouse, the changes were especially hard for women, since they would tend not to return to Göreme in the same way as their husband. Moreover, with a sharp divide between inside and outside space in the town apartments, and little or no neighbouring practices, some women who moved to Nevşehir told me they felt lonely and trapped. Yet, for many families, moving to Nevşehir where they could buy an apartment in a newly built high-rise apartment block for a fraction of the amount they received for their Göreme cave-house still seemed like the best – or sometimes only – option. During the late 2000s and early 2010s, there were so many households selling up in the old mahalles and moving to town that a dramatic population drop occurred which threatened Göreme's municipality status. Consequently, there was fear that if the population dropped to below 2000, Göreme would lose its *kasaba* status – losing with it the ability to have mayoral and council (*belediye*) governance. More over, along with there being fewer families resident in Göreme, many of those who did still live in Göreme sent their children to private school in Nevşehir thus significantly reducing the role size of Göreme's primary and intermediate schools. All in all, tourism-related displacement of residents in Göreme was becoming very literally a crisis of *"our town"*. In relation to this point, the tour agency owner quoted above continued our conversation by saying:

> The Mayor should fight for land to be released from the national park authority so that we can extend *Yeni Mahallesi* and build 500 more houses. It would be better to have all the villagers in *Yeni Mahallesi*. That would be better for everyone, then they can have their culture back, cows, sheep and so on, away from the tourists and hotels. I'm sure everyone would come back to live here, because they don't want to live away from Göreme... When I retire, what can I do in Nevşehir? I can't sit in the house all day, I can't go to the teahouse in Nevşehir, so I have to come back and live here; here I walk around and everyone says hello.

Since those who were fortunate to be able to stay in Göreme when they moved out of the increasingly touristic space of the old mahalles moved mostly to the Yeni (New) mahalle, that area became the only remaining residential neighbourhood of Göreme. While some hotels have begun to appear even in that neighbourhood in recent years, neighbours and neighbourly practices have, on the whole, continued to exist. Throughout the 2000s and 2010s, there were sofas placed outside some houses for the elderly women to sit on to do 'neighbouring' (*komşuluk*), and Abbas and Senem would still block their street to cars when making *pekmez*, during which neighbours would come and put their pot of *güveç* (stew) in the fire's ashes, staying for a while to chat and help. With a sense of the mahalle still being relatively *kapalı* (enclosed, and thus protected, safe), Abbas and Senem's front gate would often still be left slightly ajar to signal the ability for neighbours to pop in: "This is the only 'real' ma-halle now", said Senem, "the rest of Göreme has become a '*tatil köyü*'/holiday village".

Towards the latter end of the 2010s, however, new hotels were increasingly being built in Yeni Mahallesi also, so that while this mahalle still remained largely residential and a 'place of the familiar', it too was starting to become a place of tourism and the unfamiliar. In speaking to a young woman whose family had managed to stay living in Göreme, she told me that while she definitely would not like to move to an apartment in Nevşehir, as many of her friends had done, "It's also not so good here anymore – this is just a ho-tel town now", she said. This sentiment resonates with Cocola-Gant's (2018) term 'place-based displacement', discussed in Chapter 5, which refers to the feeling of 'loss of place' (p. 288) experienced by residents when the 'deep mu-tation' of the place they inhabit threatens their 'imagined future in that place' (Askland and Bunn, 2018). So, while Chapter 5 told of tourism gentrification's deep mutation of place and everyday life in Göreme's old neighbourhoods in particular, tourism-induced transformation and place-based displacement has increasingly occurred in relation to Göreme as a whole.

Having come from a young woman, the statement that "this is just a hotel town now" indicates that it is not only the older generations who experience what Albrecht (2005) termed 'solastalgia'; the 'pain experienced when there is recognition that the place where one resides and that one loves is under im-mediate assault' (Albrecht, 2005, p. 45). In its referring both to an 'erosion of the sense of belonging (identity) to a particular place' and to the 'feeling of distress…about its transformation' (ibid., p. 45), solastalgia – a sense of lost home even when still at home – has been apparent in many of the stories told throughout this book. Beyond being a crisis of control over what happens in Göreme, the crisis of "*our town*" refers to a deep ontological anxiety felt as a sense of lost home. The feeling of distress about Göreme's transformation was captured most acutely in Ali's exclaiming, "Even if everybody's making good money… *we can never buy Göreme back!*". Highly emotionally charged, this ex-clamation encapsulates the sense of there being, folded in together, a crisis of "our town" and an all-encompassing ontological crisis, resonating once more with the notion of the spectre of unlimited change. It is little wonder that cries of "Enough!/*Yeter*!" increasingly ring out across Göreme.

The spectre of unlimited change

In depicting the stories and major strands of Göreme's tourism and social change over four decades as culminating in three crises, I am aware of the need to be cautious both of my own pessimism, and my own attempts at empathy: (How) can I feel loss or solastalgia *for* and *with* Neriman who finds herself living in Aydınlı Mahallesi surrounded by hotels and without neighbours?; and (how) can I feel *hüzün* (melancholy) *for* and *with* Abbas when he explains why he built a fence around his garden and he tells me his grapevines cannot breathe? Inevitably, when rendering crisis narratives from stories of change, a certain future is imagined, which necessitates calling my own narrative authorisation into question. That said, naming the three crises *as crises* is to recognise the synergies that Göreme's story, or stories, of tourism and social change have with the great many places around the world that have experienced the sense of accelerated change and runaway processes which Eriksen (2016) argues have, since the turn of the millennium, become the reality of the contemporary neoliberal world *everywhere*. In other words, the three crises I have named *as crises* might be considered crises of our times: Just after the turn of the millennium Mireille Rosello argued that 'hospitality is indeed in crisis, not simply because our contemporary (Western?) world may not have enough of it, but because it is in the process of being redefined' (2001, p. 8); Diane Coyle (2011) addressed the 'crisis of enough' in her book '*The Economics of Enough: How to Run the Economy as If the Future Matters*'; and in their writing about the concept of solastalgia, Asklund and Bunn (2018) suggest that place-based distress – experienced in what I have called the crisis of "our town" – is a deeply ontological crisis capable of utterly disrupting people's ability to imagine the future. An inability to imagine the future seems very much aligned with the notion of the spectre of unlimited change. It is the *unlimited* prospect of change – change which will render life utterly unrecognisable from what it was before – that appears to have generated these new instabilities which Eriksen (2016) identifies as the 'overheating' characteristic of the contemporary neoliberal world.

Yet, my long-term involvement in Göreme nonetheless affords me an ability to see that the three crises outlined above – of "our town", of hospitality, and of enough – are in many ways ambivalent crises, as perhaps all crises are: "Yes, people are getting rich, we are making good money, but at the same time *we are losing Göreme ... we can never buy Göreme back!*" While crises, as imagined, are apocalyptic and usually dystopian, crises – like spectres – inevitably hang in the air differently for different generations and, as such, may simultaneously be utopian in their creating possibilities of the not-yet-become. My longitudinal ethnography has thus enabled understanding not only of how the crises, and the spectre of unlimited change, have developed, but also how different generations of peasants-in-transition negotiate their positioning within the broader tourism and social change nexus. As I argued in Chapters 6 and 7, for example, the frontiers of change can be viewed as opening up new spaces of possibility to re-work what is 'doable' for girls and young women. While showing social change to always be relational, contingent, and ambivalent, such a highlighting of the 'potential for the reworking, disruption, contestation, transgression and transformation of the dominant codes and behaviours of society' (Aitchison, 2005, p. 217) puts an altogether more hopeful slant on the prospect of unlimited change than referring to that prospect as a spectre.

Indeed, it is always tempting to emphasise, or even to celebrate, such instances of hope-filled local agency and resistance, not only to patriarchal dominances but also to the imagined, or actual, seemingly bulldozer-like forces of global neoliberal capitalism – in this instance, in the form of tourism. This is why, while keeping in sight the sticky shame that can often accompany associations with '*köylülük*' (peasantry) and with being *kapalı* (enclosed, covered), it is something of a comfort that, in line with Turkey's 'new peasantry' more generally (Öztürk et al., 2018), the people of Göreme appear to continue to value their gardens, as well as the internal communing ethos of their remaining neighbourhoods. It is also why I enjoy observing and writing about my observations of the continued valuing of non-commodity relationships and reciprocity in Göreme's tourism; these I see as examples of the adaptive continuation of peasant moral economy and sociality which, as Öztürk et al. (2018) suggest, might manifest as a kind of resistance to 'the neoliberal squeeze of capital' (ibid., p. 245). After all, it is such a 'clear-eyed engagement with notions of limits' that, according to Higgins-Desboilles (2018, p. 159), 'the current culture of consumerism and pro-growth ideology precludes'. Indeed, against a growing realisation of the failings of 'the predominant view of the First World that "natural resources" are unlimited' (Nash, 2007, p. 36), peasant and other cultural notions of limits and the related need for balance are increasingly called upon to provide hopeful alternatives. As Deborah Bryceson (2000) suggested: 'The social and political legacy of peasant culture is bound to endure long after peasant economies disappear... There is even the possibility that future global...crises will be resolved by drawing on the world's peasant legacy' (p. 324).

This is precisely where the notions of *limited* good and *unlimited* good come to prominence in what my Göreme stories can tell us about living with tourism and about social change more broadly in the contemporary neoliberal world. On the one hand, by attending to a multiplicity of everyday stories, it is apparent that the experience, and the doing, of social change is always ambivalent due to its ever-proliferating – and often clashing and contradictory – affordances, enactments, and affects. So, while referring to the metaphor of a frontier is useful – that is, in its conjuring of quirky and unpredictable non-linear leaps and skirmishes which proliferate as the frontier of change (Tsing, 2005) – I find the ever-presence of ambivalence to be the most productive depiction of how social change manifests as incoherencies in the contemporary neoliberal world. This ever-present ambivalence locates as a space between; between sticky memories and paradoxical hopes and dreams, between the intentional and unintentional 'unleashings' of a multiplicity of actors – human and non-human, balloon and non-balloon, and of course, between the daily practices of sociality borne out of an image of limited good and the seemingly limitless possibilities promised by a world of unlimited good. Just as Qian and Wei (2020) have concluded from their recent observations in a rural tourism locale in Southwest China that 'Non-capitalist elements are not capital's others, [but rather], capitalism is always-already a capital–non-capital complex' (p. 251), so, I would argue, living with tourism is always-already a *limited good–unlimited good* complex. My Göreme stories have shown that this complex inevitably arises as the people in tourism places become increasingly embedded in global tourism market relations, along with other global trends and fluxes, while they also maintain their embeddedness in locally familiar relations and practices. In Göreme, as everywhere perhaps, locally embedded tourism practices and relations continue, in an always-ongoing negotiation, to form fissures in global capitalism's apparent indifference.

On the other hand, my longitudinal ethnographic study in Göreme has afforded insights into how it is that global neoliberal capitalism's prospect of unlimited good has come to be so pervasive. As I remarked in Chapter 1 of this book, at the turn of the millennium, I would not have predicted that within two decades, Göreme would become the hot-air-ballooning-tourism centre of the world, nor that ballooning there would become so valuable in the global social media 'like economy', nor that, consequently, tourism and tourists – out of a desire to take selfies with the balloons and with Göreme merely as a backdrop – would become such an apparently unlimited, yet certainly ambivalent, "good" for Göreme. I would not have predicted that Göreme's old neighbourhoods would become so emptied of neighbours, nor that so many orchards and vineyards would be stripped of their vines and apricot trees to be turned into hot-air balloon take-off sites. I had no foresight, in other words, of the extent to which the ability of Göreme's peasants-in-transition to resist and disrupt the 'neoliberal squeeze of capital' would lessen so that the neoliberal squeeze of capital would indeed come to disrupt and transform the non-commodity relationships and other aspects of sociality which had previously, necessarily, hinged on an image of limited good. While practices and relations of daily life, including tourism business-related practices, might always oscillate between limited good principles and unlimited good aspirations, along with affording new hopes and dreams for many the 'unlimited' prospect – of 'good' and of change – has also generated new instabilities and indeterminacies. The everyday thus becomes a deeply ambivalent site of affective negotiation. The spectre of unlimited change conjures these always-ambivalent dynamics as well as the always-ambiguous ontological tensions manifested when living life within limits gives way to a prospect of unlimited good, and unlimited change.

Post-Covid postscript

I have found during visits to Göreme in recent years that I tend to evoke a nostalgic urge to reminisce. This was especially so when I visited in the summer of 2022, after a three-year-long period of not being able to travel to Göreme because of Covid-19-related border closures. The Covid-19 pandemic had passed in Göreme similarly to many other places around the world; incurring periods of lockdown, severely restricted mobility, job losses – especially in the tourism and hospitality sectors – accompanied by wage subsidies to prop up livelihoods, and untimely deaths. The pandemic also created a time to pause and reflect, about how overheated things had become just prior to Covid-19 and whether things should resume that same path once the pandemic is over. During the pandemic, many people had so enjoyed the sense of a return to "how things used to be" that they feared tourism's return. Such fears were prompted at a global level too, with Covid-19 'widely seen as a potential moment of transformation', 'a possible game changer for globalisation as well as for global tourism' (Higgins-Desbiolles, Bigby and Doering, 2022). Against a pre-pandemic context of so-called 'overtourism' in many locales around the world, the pandemic gave rise to – perhaps overly hope-filled – beliefs that a reversing, or even an undoing, of things might be possible. Along with calls to 'socialise tourism' in order that tourism might come to better serve the wellbeing of communities rather than the other way around (ibid.), terms such as 'degrowth' and 'regenerative tourism', and notions of planetary limitations even, surged, or resurged, to prominence.

Yet, when I visited Göreme in 2022, things were showing signs of picking up bigger and stronger than ever; many new businesses had opened and all were poised ready for tourists' big return. People knew that the pause was just a pause and that tourism would soon return to 2018–19 levels, or greater. After all, tourism in Göreme had experienced many such pauses before – usually due to political troubles in Turkey or beyond – and had always bounced back seemingly unscathed. Knowing that Covid-19 was merely a temporary reprieve prompted both nostalgic and solastalgic yearnings for many. For instance, Abbas told me that he spent much time during the period of the pandemic reflecting on how, as a child, he used to play in the streets in his old mahalle: "How I miss those days", he said, whereas now *"turistler geldi* (tourists came), *para geldi* (money came), *insanlık bitti* (humanity has finished)". As if directly referring to the rupture experience of solastalgia, in a melancholic tone he added: "Tourism has broken everything". Several other people similarly told me that Göreme, and Göremeli people, are now "broken": "Göreme is not liveable anymore", a friend – a woman in her forties and the wife of one of Göreme's more successful businessmen – told me. As well as being noisy and having too much traffic, "it is just a business place now", she said, "a place for making money, not for living". She continued by saying that having so much money has driven the Göremeli people crazy: "They don't know what to do with all this money!"

Other women I spoke to during my post-Covid visit – Zekiye and others who run craft stalls on Canal Street, Hanife and Hayriye who still run their shop on the main street selling locally crafted jewellery, and Filiz who had been able to hire just one Afghani worker to help her restore her cave-house in Aydınlı Mahallesi – also reminisced about how nice and quiet Göreme was during the pandemic, with no balloons, no ATVs, and people returning to tend their gardens. However, while Covid-19 had provided a taste of "before-tourism", these women's nostalgia tended to be of the more future-oriented, hopeful variety I discussed in Chapter 7, wherein they reiterated a desire *not to be* who they once were. Zekiye talked again of the *özgürlük* (freedom) running the stalls affords her and the other stall-holder women. Filiz had finished converting her house into a small hotel and told me that she likes being the *'patron'* (boss), although she finds it tiring always having to please the guests even when she does not feel like it; I watched on as she ably answered her guests' questions about how to make a balloon-ride booking and what time their airport shuttle would pick them up. Hanife and Hayriye told me about their going to Istanbul earlier in the year to attend a trade fair; they felt a bit "shy" at first about going there, but they encouraged each other and the trip was successful in securing some new commission contracts, in turn allowing them to employ even larger numbers of women in "poor villages".

I also had a chance during my 2022 visit to catch up with all three Mehmets to whom I have referred in this book. Similarly to others' reminiscing about the quietness during the pandemic, 'Walking Mehmet' remarked that "Covid-19 was a gift from nature to nature": Everything stopped, he said, there were no tourists, no cars, no balloons, and the plants and trees could breathe again. A stork had even made a nest and raised a chick at the top of the fairy chimney behind his house, he told me; "It was the first time I've seen this in years!" I asked him why he no longer kept chickens in the courtyard behind his house – he had been very proud of the chicken coop he had built when I had seen him in 2019 - and he told me that when the hotel next door had resumed business again after the pandemic, the owners had complained that the chickens' stench and the cockerel's crowing disturbed their guests. Meanwhile, Mehmet who was introduced earlier in this chapter compared the quietness during the pandemic with how things were becoming now that tourism was returning. A new restaurant-bar had opened next door from which the thumping music kept him awake until the early hours of each morning. He also talked again about how, when he walks up the mainstreet between his home and the men's teahouse, nobody knows him and the restaurant touts try to lure him in to buy a drink: "I hate it, I don't feel I belong in my own town any more", he said. The third Mehmet, introduced in Chapter 2 in relation to his *Instagram*-influencer work, said that the pandemic had been the best time of his life. Not having to do much of the usual work related to his multiple businesses, he was able to have plenty of family time and go for walks every day in the valleys with his son. This reminiscing led him to reflect on how much Göreme has been changing; how it is "not a place for living in anymore", and how even Yeni Mahallesi is becoming filled with hotels. I asked him why, if he thinks this way, does he continue to work on social media in order to keep on promoting Göreme. After pausing for a few seconds to collect his thoughts on how to answer, he replied that after Covid-19, they needed to build tourism back up again – they needed to ensure that Göreme stays prominent in the global marketplace; that – especially with the Russia–Ukraine war, and China's border still closed at the time – they needed to make sure that Göreme was known to other markets around the world so that it could fully recover from the pandemic-pause.

Along with Mehmet, several of Göreme's prominent businessmen had used the pandemic-pause to progress some of the Göreme-wide projects which they had been planning before Covid-19. They established a 'social enterprise' organisation which set up a women's cooperative lunch restaurant – serving everyday local food rather than 'tourist food' – as well as doing baking and making local dishes for other Göreme cafés and restaurants. The cooperative also had contracts with the region's hot-air ballooning companies to provide 'breakfast bags' – a carton of juice, an apple, biscuits, and nuts – for their flight customers. The women busy working in the 'breakfast bag' section of the cooperative building told me they produced and distributed two-and-a-half thousand such bags each day. Another of the social enterprise's main projects was to establish a Göremeli-owned online booking agency; with the global online booking agencies earning several million U.S. dollars in revenue from Göreme's hotels each year, the Göremeli businessmen working on the project expected that if they could retain just one million dollars through their own booking agency, they would be able to use the money for other social and environmental projects, such as childcare to enable women to work, children's clubs and school-holiday activities, preservation work in the old mahalles, and environmental conservation work in the valleys. Another project – a culture-preservation endeavour – was underway in a location several kilometres from Göreme; this one was entirely funded by Ali who was introduced in Chapter 7. This multimillion-dollar project was aimed at creating a school-cum-village where both children and adults could come to learn the traditional organic farming techniques, regional food production, and preservation methods – no doubt including *pekmez*-making – and other regional cultural practices.

Meanwhile, certain 'projects' enacted at a government level were also underway. The disestablishment – by Presidential decree – of the Göreme National Park in late 2019 had paved the way for the newly established Cappadocia Area Directorate to continue its campaign to rid the area of "illegally-built structures", such as wooden café-stalls set up by villagers long ago to provide drinks-stops for tourists hiking in the valleys. In 2021, the directorate decreed that a new road would be constructed between Göreme and Ortahisar. While the new road was ostensibly built to replace the old one that was deemed too close – and therefore causing damage to – the Göreme Open-Air Museum, making way for the new road reportedly involved knocking down other 'fairy-chimney' areas of historic value. Amidst significant media-covered protestations, the new road was completed and opened in late 2022. The replacement of National Park status and governance with the Cappadocia Area Directorate has also caused consternation regarding the expectation that the new presidential-directorate was opening the door for big business – including major construction companies – to swoop in and build mega-projects. Such a mega-project already underway when I visited in 2022 was the construction, in nearby Ortahisar, of an adventure-theme park, with large hotels being constructed nearby to accommodate visitors to the theme park. Promoted as a "fairytale full of adventure", the website of the company constructing the theme park clearly illustrates the 'image of unlimited good' view of tourist numbers, or 'markets', as expandable: 'The unique atmosphere of Cappadocia, which already attracts tens of thousands of tourists every year, will host *thousands more* with this adventure park, which will add value to the region throughout the year' (excapturkey.com, 2021). No doubt, with the disestablishment of the Göreme National Park, such examples of neoliberal capitalism will become increasingly plentiful there in years to come. Indeed, in 2022, a quick tour around Göreme's surrounding valleys showed that the entire area has become akin to an *Instagram* theme park, with tethered hot-air balloons, lavender gardens, the ubiquitous heart-shaped frames, and even giant cranes erected to allow photos to be taken whilst swinging out over fairy-chimney-filled valleys. If Covid-19 was a time to pause and reflect, it is fascinating to ponder where all of these entangled 'tourism orderings' – and non-orderings – surrounding contemporary tourists' seemingly relentless quest to capture on their smartphones the extraordinary perfection in themselves will go next.

Indeed, it did not come as much of a surprise when I was told during my 2022 visit that Göreme is "at a tipping point"; whilst big external companies position themselves to construct their mega-projects, Göremeli business owners scurry to protect their position as – what they consider to be – the rightful owners, earners, and decision-makers vis-à-vis Göreme's tourism. When I asked Mustafa, the owner of Aydınlı Cave-Hotel, who had made such a "tipping point" comment, why – having just purchased a second hotel – he felt the need to grow his business, he answered that whenever he and his Göremeli peers had the money to do so, they would buy hotels and other property in order to "block outsiders" from buying them. During our "tipping point" conversation, Mustafa walked me through the stone-cut corridors of his Aydınlı Cave-Hotel, on the walls of which were displayed several photographs showing how Aydınlı Mahallesi looked two to three decades ago. Looking at the photos, he talked about how the mahalle used to be full of life with all the neighbours (*komşuler*) and neighbouring (*komşuluk*), making *pekmez* and bread together and sharing a "real" social life: "Now I feel this is not my village any more – I feel very very bad", he said, this statement a reminder once again of Rosaldo's (1989) earlier observation that nostalgic yearning is, perhaps all too often, a mourning for what one has destroyed.

As I finish this book manuscript in early 2023 – 20 years on from the year in which *Living with Tourism (1)* was published, I receive a message from a friend in Göreme with a photo attached of a large Buddha statue that had just been erected in front of a new restaurant on Göreme's mainstreet. The photo raises yet further uncertainty regarding where all the troubled and troubling stories told throughout this book will lead as Göreme moves further not only into this 'post-pandemic' decade but also more fully into the uncertainties of global neoliberalisation. Just as the big balloons had embodied the spectre of unlimited change 20 years ago, although we did not see it at the time, the big Buddha and the Big Bus would appear to do so today, and this time, we do see it. Yet so too, in an altogether more hopeful way, which is why this time we *must* see it, does Neriman's granddaughter, in her heading off to university, to study law. I therefore wonder what future possibilities the stories told here open up, as well as what futures the stories foreclose. Indeed, the first half of 2023 also saw a devastating earthquake in the southeastern part of Turkey, prompting significant humanitarian relief effort from Cappadocia, including many of the region's hotels being used to house those the earthquake had displaced, and in May'23 Turkey's polarising general election took place. These events serve as yet another reminder – as the Covid-19 pandemic itself had done – that Göreme, as everywhere, is always-already entangled with everywhere else. Yet, rather than blaming externalities, it needs also to be remembered that, with Göreme's tourism business continuing to be largely under local ownership, the foreboding spectre of unlimited change is, in many ways, a spectre of dreams come true; such is the paradox of continuing to drive the runaway car even though its brakes have long shown signs of failing. I wonder, if I or somebody else were to write a book about Göreme in another 20 years' time, what stories will there be about *Living with Tourism* then? Perhaps only the spectre of time can tell.

Bibliography

Abu-Lughod, L. (1986) *Veiled Sentiments: Honor and Poetry in a Bedouin Society.* Berkeley: University of California Press.

Adams, K. (2006) *Art as Politics: Re-crafting Identities, Tourism, and Power in Tana Toraja, Indonesia.* Honolulu: University of Hawai'i Press.

Adams, K. (2016) Tourism and ethnicity in insular Southeast Asia: Eating, praying, loving and beyond. *Asian Journal of Tourism Research*, 1 (1): 1–28.

Ahmed, S. (2004) Affective economies. *Social Text*, 79 (2): 117–139.

Ahmed, S. (2014) *The Cultural Politics of Emotion.* London: Routledge.

Aitchison, C. (2005) Feminist and gender perspectives in tourism studies: The socio-cultural nexus of critical and cultural theories. *Tourist Studies*, 5 (3): 207–224.

Albrecht, G. (2005) 'Solastalgia'. A new concept in health and identity. *PAN: Philosophy Activism Nature*, 3: 41–55.

Alp, A. (2019) Protection unit proposed for Cappadocia, Hürriyet Daily News. Retrieved from: www.hurriyetdailynews.com/protection-unit-proposed-for-cappadocia-143286 [09/05/2019].

Anderson, B. (2006) Becoming and being hopeful: Towards a theory of affect. *Environment and Planning D: Society and Space*, 24: 733–752.

Ap, J., and Wong, K.K.F. (2001) Case study on tour guiding: Professionalism, issues, and problems. *Tourism Management*, 22: 551–563.

Araghi, F. (1995) Global depeasantisation, 1945–1990. *Sociological Quarterly*, 36(2): 337–368.

Arin, C. (2001) Femicide in the name of honor in Turkey. *Violence Against Women*, 7: 821–825.

Askland, H.H., and Bunn, M. (2018) Lived experiences of environmental change: Solastalgia, power and place. *Emotion, Space and Society*, 27: 16–22.

Aslaner, M.A. (2019) Hot-air balloon tour accidents in the Cappadocia region. *Aerospace Medicine and Human Performance*, 90 (2): 123–127. Retrieved from: https://pubmed.ncbi.nlm.nih.gov/30670122 [3/3/2021].

Bærenholdt, J., Haldrup, M., Larsen, J., and Urry, J. (2004) *Performing Tourist Places.* Aldershot: Ashgate.

Bailey, F.G. (1969) *Strategems and Spoils – The Social Anthropology of Politics.* Oxford: Basil Blackwell.

Bailey, F.G. (1971) Gifts and poison, in F.G. Bailey (ed.) *Gifts and Poison – The Politics of Reputation* (pp.1–26). Oxford: Basil Blackwell.

Bell, D. (2007) Moments of hospitality, in J.G. Molz and S. Gibson (eds.) *Mobilizing Hospitality: The Ethics of Social Relations in a Mobile World* (pp. 29–46). Aldershot: Ashgate.

Benali, A., and C. Ren (2019) Lice work: Non-human trajectories in volunteer tourism. *Tourist Studies*, 19 (2): 238–257.

Bezmen, C. (1996) *Tourism and Islam in Cappadocia*, unpublished Ph.D. thesis, Cambridge University, Cambridge.

Bhabha, H. (1984) Of mimicry and man: The ambivalence of colonial discourse. *JSTOR*, 28: 125–133.

Bloch, N. (2021) *Encounters across Difference: Tourism and Overcoming Subalterity in India*. London: Lexington Books.

Bourdieu, P. (1965) The sentiment of honour in Kabyle society, in J.G. Peristiany (ed.) *Honour and Shame – The Values of Mediterranean Society* (pp. 191–242). London: Weidenfeld and Nicolson.

Boym, S. (2001) *The Future of Nostalgia*. New York: Basic Books.

Bradbury, J. (2012) Narrative possibilities of the past for the future: Nostalgia and hope. Peace and Conflict: Journal of Peace Psychology, 18 (3): 341–350.

Brouder, P., Clave, S., Gill, A., and Ioannides, D. (2017) Why is tourism not an evolutionary science? Understanding the past, present and future of destination evolution, in P. Brouder, S. Clave, A. Gill, and D. Ioannides (eds.) *Tourism Destination Evolution* (pp. 1–18). Abingdon: Routledge.

Brouder, P., Clave, S., Gill, A., and Ioannides, D. (eds.) (2017) *Tourism Destination Evolution*. Abingdon: Routledge.

Bruner, E. (1986) Introduction: Experience and its expressions, in V. Turner and E. Bruner (eds.) *The Anthropology of Experience* (pp. 3–30). Urbana: University of Illinois Press.

Bryceson, D. (2000) Disappearing peasantries? Rural labour redundancy in the neoliberal era and beyond, in D. Bryceson, C. Kay and J. Moooij (eds.) *Disappearing Peasantries?: Rural Labour in Africa, Asia and Latin America* (pp. 299–326). London: ITDG Publishing.

Butler, R.W. (1980) The concept of a tourist area cycle of evolution: Implications for management of resources. *The Canadian Geographer*, 24 (1): 5–12.

Christou, P., Farmaki, A., Saveriades, A., and Georgiou, M. (2020) Travel selfies on social networks, narcissism and the "attraction-shading effect". *Journal of Hospitality and Tourism Management*, 43: 289–293.

Cocola-Gant, A. (2018) Tourism gentrification, in L. Lees and M. Phillips (eds.) *Handbook of Gentrification Studies* (pp. 281–293). Cheltenham: Edward Elgar Publishing.

Cole, S. (2008) *Tourism, Culture and Development: Hopes, Dreams and Realities in East Indonesia*. Clevedon: Channel View Publications.

Cone, C.A. (1995) Crafting selves: The lives of two Mayan women. *Annals of Tourism Research*, 22 (2): 314–327.

Coyle, D. (2011) *The Economics of Enough: How to Run the Economy as If the Future Matters*. Princeton and Oxford: Princeton University Press.

Crawshaw, C., and Urry, J. (1997) Tourism and the photographic eye, in C. Rojek and J. Urry (eds.) *Touring Cultures: Transformations of Travel and Theory* (pp. 177–195). London: Routledge.

Crick, M. (1994) *Resplendent Sites, Discordant Voices: Sri Lankans and International Tourism*. Switzerland: Harwood Academic Publishers.

Dawney, L. (2011) The motor of being: A response to Steve Pile's 'emotions and affect in recent human geography'. *Transactions of the Institute of British Geographers*, 36 (4): 599–602.

Delaney, C. (1991) *The Seed and the Soil – Gender and Cosmology in Turkish Village Society*. Berkeley and Los Angeles: University of California Press.

Delaney, C. (1993) Traditional modes of authority and co-operation, in P. Stirling (ed.) *Culture and Economy – Changes in Turkish Villages* (pp. 140–155). Cambridgeshire: The Eothen Press.

Derrida, J. (1999) *Adieu to Emmanuel Levinas*. Translated by P.A. Brault and M. Naas. Stanford: Stanford University Press. (Originally published as Adieu á Emmanuel Levinas, Paris: Galilée, 1997).

Dinhopl, A., and Gretzel, U. (2016) Selfie-taking as touristic looking. *Annals of Tourism Research*, 57: 126–139.

Doucet, B. (2009) Living through gentrification: Subjective experiences of local, non-gentrifying residents in Leith, Edinburgh. *Journal of Housing and the Built Environment*, 24: 299–315.

Dundes, A. (ed.). (1992) *The Evil Eye: A Casebook*. Madison: University of Wisconsin Press.

Durakbaşa, A., and Karapehlivan, F. (2018) Progress and pitfalls in women's education in Turkey (1839–2017). *Encounters in Theory and History of Education*, 19: 70–89.

Edensor, T. (1998) *Tourists at the Taj: Performance and Meaning at a Symbolic Site*. London: Routledge.

Edensor, T. (2006) Sensing tourist spaces, in C. Minca and T. Oakes (eds.) *Travels in Paradox* (pp. 23–45). Lanham, MD: Rowman and Littlefield.

Egresi, I. (2016) Tourism and sustainability in Turkey: Negative impact of mass tourism development, in I. Egresi (ed.) *Alternative Tourism in Turkey: Role, Potential Development and Sustainability* (pp. 35–53). Cham, Switzerland: Springer International Publishing.

Elmas, S. (2007) Gender and tourism development: A case study of the Cappadocia Region of Turkey, in A. Pritchard (ed.) *Tourism and Gender: Embodiment, Sensuality and Experience* (pp. 302–314). Wallingford: CAB International.

Elsrud, T. (2001) Risk creation in travelling: Backpacker adventure narration. *Annals of Tourism Research*, 28 (3): 597–617.

Emge, A. (1992) *Change in Traditional Habitat, Traditional Dwellings and Settlements Working Paper Series* (Vol. 37). Berkeley, CA: Centre for Environmental Design Research.

Erbil, Ö. (2017) Deadly balloon crashes in Turkeys Cappadocia, Hurriyet Daily News (April, 2017). Retrieved from: Hurriyetdailynews.com/deadly-balloon-crashes-in-Turkeys-Cappadocia-turn-eyes-to-malfunctions-idle-tower-111923 [3/3/2021].

Eriksen, T.H. (2016) *Overheating: An Anthropology of Accelerated Change*. London: Pluto Press.

Erman, T. (2016) *Mış gibi site: Ankara'da Bir TOKI-Gecekondu Dönüsüm Sitesi [Like a Housing Estate: A TOKI-Gecekondu Transformation Housing Estate in Ankara]*. Istanbul: Iletişim Yayınları.

Everett, S. (2016) *Food and Drink Tourism: Principles and Practice*. London: Sage.

Everingham, P. (2016) Hopeful possibilities in spaces of 'the-not-yet become': Relational encounters in volunteer tourism. *Tourism Geographies*, 18 (5): 520–538.

Everingham, P., Obrador, P., and Tucker, H. (2021) Trajectories of embodiment in Tourist Studies, *Tourist Studies*, 21 (1): 70–83.

Excapturkey. (2021) Retrieved from: https://www.excapturkey.com/en [15/08/2022].

Foster, G. (1965) Peasant society and the image of limited good. *American Anthropologist*, 67 (2): 293–315.

Foster, G. (1966) Foster's reply to Kaplan, Saler and Bennett. *American Anthropologist*, 68 (1): 210–214.

Foster, G. (1972) A second look at limited good. *Anthropological Quarterly*, 45 (2): 57–64.

Franklin, A. (2012) The choreography of a mobile world: Tourism orderings, in R. van der Duim, C. Ren, and G.T. Jóhannesson (eds.) *Actor-Network Theory and Tourism: Ordering, Materiality and Multiplicity* (pp. 43–58). London: Routledge.

Franklin, A. (2014) On why we dig the beach: Tracing the subjects and objects of the bucket and spade for a relational materialist theory of the beach. *Tourist Studies*, 14 (3): 261–285.

Franklin, A., and Crang, M. (2001) The trouble with tourism and travel theory. *Tourist Studies*, 1 (1): 5–22.

Garbutt, R.G. (2006) The locals: A critical survey of the idea in recent Australian scholarly writing. *Australian Folklore*, 21: 172–192.

Gedalof, I. (2003) Taking (a) place: Female embodiment and the re-grounding of community, in S. Ahmed, C. Castaneda, A.M. Fortier, and M. Sheller (eds.) *Uprooting/Regroundings: Questions of Home and Migration* (pp. 91–114). Oxford: Berg.

Gerlitz, C., and Helmond, A. (2013) The like economy: Social buttons and the data-intensive web. *New Media & Society*, 15 (8): 1348–1365.

Germann Molz, J. (2018) Discurses of scale in network hospitality: From the Airbnb home to the global imaginary of "belong anywhere. *Hospitality and Society*, 8 (3): 229–251.

Germann Molz, J., and Paris, C.M. (2015) The social affordances of flashpacking: Exploring the mobility nexus of travel and communication. *Mobilities*, 10 (2): 173–192.

Gibson, J.J. (1979) The theory of affordances, in J.J. Gieseking, W. Mangold, C. Katz, et al. (eds.) *The People, Place, and Space Reader* (pp. 56–60). Abingdon: Routledge.

Giddens, A. (2003) *Runaway World: How Globalization Is Reshaping Our Lives*. New York: Routledge.

Gilmore, D. (ed.) (1987) *Honour and Shame and the Unity of the Mediterranean*. Washington, DC: American Anthropological Association.

Goddard, V. (1989) Honour and shame: The control of women's sexuality and group identity in Naples, in P. Caplan (ed.) *The Cultural Construction of Sexuality* (pp. 166–192). London: Routledge.

Goodwin, H. (2017) The challenge of overtourism. Responsible Tourism Partnership Working Paper 4, October 2017. Retrieved from: https://haroldgoodwin.info/pubs/RTP'WP4Overtourism01'2017.pdf. [22/07/2021].

Gravari-Barbas, M., and Guinand, S. (2017) Introduction, in M. Gravari-Barbas and S. Guinand (eds.) *Tourism and Gentrification in Contemporary Metropolises* (pp. 1–21). London: Routledge.

Grit, A. (2014) Messing around with serendipities, in S. Veijola, J. Germann Molz, O. Pyyhtinen, E. Höckert, and A. Grit (eds.) *Disruptive Tourism and Its Untidy Guests: Alternative Ontologies for Future Hospitalities* (pp. 122–141). London and New York: Palgrave Macmillan.

Guyer, G., Lambin, E., Cliggett, L., Walker, P., Amanor, K., Bassett, T., Colson, E., Hay, R., Homewood, K., Linares, O., Pabi, O., Peters, P., Scudder, T., Turner, M., and Unruh, J. (2007) Temporal heterogeneity in the study of African land use: Interdisciplinary collaboration between anthropology, human geography and remote sensing. *Human Ecology*, 35: 3–17.

Hall, C.M., Mitchell, R., and Sharples, L. (2003) Consuming places: The role of food, wine and tourism in regional development, in C.M. Hall, L. Sharples, R. Mitchell, N. Macionis, and B. Cambourne (eds.) *Food Tourism around the World: Development Management and Markets* (pp. 25–59). Oxford: Butterworth Heinemann.

Hall, M., and Williams, A. (2008) *Tourism and Innovation*. London: Routledge.

Hannerz, U. (1990) Cosmopolitans and locals in world culture, in M. Featherstone (ed.) *Global Culture: Nationalism, Globalisation and Modernity* (pp. 237–251). London: Sage.

Haraway, D. (2003) The science question in feminism and the privilege of partial perspective, in Y. S. Lincoln and N. K. Denzin (eds.) *Turning Points in Qualitative Research: Tying knots in a handkerchief* (pp. 21–46). Walnut Creek, CA: Altamira Press.

Harrison, J. (2003) *Being a Tourist: Finding Meaning in Pleasure Travel*. Vancouver: University of British Columbia Press.

Harrison, J. (2019) "So in effect I was studying myself": Knowing (our) tourist stories, in N.M. Leite, Q.E. Castaneda, and K.M. Adams (eds.) *The Ethnography of Tourism: Edward Bruner and Beyond* (pp. 71–85). London: Lexington Books.

Hartal, G. (2020) Touring and obscuring: How sensual, embodied and haptic gay touristic practices construct the geopolitics of pinkwashing. *Social & Cultural Geography*, 23 (6): 836–854.

Heal, F. (1990) *Hospitality in Early Modern England*. Oxford: Clarendon Press.

Herzfeld, M. (1987) "As in your own house": Hospitality, ethnography, and the stereotype of Mediterranean Society, in D. Gilmore (ed.) *Honour and Shame and the Unity of the Mediterranean* (pp. 75–89). Washington, DC: American Anthropological Association.

Herzfeld, M. (1980) Honour and shame: Problems in the comparative analysis of moral systems. *Man*, 15: 339–351.

Hiebert, D. (2002) Cosmopolitanism at the local level: The development of transnational neighbourhoods, in S. Vertovec and R. Cohen (eds.) *Conceiving Cosmopolitanism: Theory, Context and Practice* (pp. 209–233). Oxford: Oxford University Press.

Higgins-Desboilles, F. (2018) Sustainable Tourism: Sustaining tourism or something more? *Tourism Management Perspectives*, 25: 157–160.

Higgins-Desboilles, F., Bigby, B.C., and Doering, A. (2022) Socialising tourism after COVID-19: Reclaiming tourism as a social force? *Journal of Tourism Futures*, 8 (2): 208–219.

Hovardaoğlu, O., and Çalişir-Hovardaoğlu, S. (2021) Uneven transformation of traditional agricultural producers into hybrid peasant-entrepreneurs through social media. *The Journal of Rural and Community Development*, 16 (1): 86–107.

Howell, S., and Talle, A. (2012) Introduction, in S. Howell and A. Talle (eds.) *Returns to the Field* (pp. 1–22). Bloomington: Indiana University Press.

Hurriyet Daily News. (2019) Change in Göreme valley status to prevent illegal constructions (October 22 2019). Retrieved from: https://www.hurriyetdailynews.com/change-in-Göreme-valley-status-to-prevent-illegal-constructions-ministry-147823 [29/09/2022].

Hviding, E. (2012) Compressed globalization and expanding desires in Marovo Lagoon, Solomon Islands, in S. Howell and A. Talle (eds.) *Returns to the Field* (pp. 203–229). Bloomington: Indiana University Press.

Ingold, T. (2011) *Being Alive: Essays on Movement, Knowledge and Description.* London: Routledge.

Jiménez-Esquinas, G. (2017) "This is not only about culture": On tourism, gender stereotypes and other affective fluxes. *Journal of Sustainable Tourism*, 25 (3): 311–326.

Jonah. (2018) Retrieved from: Filmschoolrejects.com/short-film-jonah-is-a-chaotic-gorgeous-fish-story-fd6aa84e2492 [21/3/2022].

Jóhannesson, G.T., and Lund, A.L. (2019) Beyond overtourism: Studying the entanglements of society and tourism in Iceland, in C. Milano, J. Cheer, and M. Novelli (eds.) *Overtourism: Excesses, Discontents and Measures in Travel and Tourism* (pp. 91–106). Wallingford: CABI.

Jóhannesson, G.T., Ren, C., and van der Duim, R. (2012) Gatherings: Ordering, materiality and multiplicity, in R. van der Duim, C. Ren, and G.T. Jóhannesson (eds.) *Actor-Network Theory and Tourism: Ordering, Materiality and Multiplicity* (pp. 164–173). London: Routledge.

Jóhannesson, G.T., Ren, C., and van der Duim, R. (2016) ANT, tourism and situated globality: Looking down in the Anthropocene, in M. Gren and E. Huijbens (eds.) *Tourism and the Anthropocene* (pp. 77–93). London: Routledge.

Johnson, H. (2004) Subsistence and control: The persistence of the peasantry in the developing world. *Undercurrent*, 1: 55–65.

Jokenen, E., and Veijola, S. (2003) Mountains and landscapes: Towards embodied visualities, in D. Crouch and N. Lubbren (eds.) *Visual Culture and Tourism* (pp. 259–278). Oxford: Berg.

Jordan, S. (2019) *Postdigital Storytelling: Poetics, Praxis, Research.* London: Routledge.

Kandiyoti, D. (2002) Introduction: Reading the fragments, in D. Kaniyoti and A. Saktanber (eds.) *Fragments of Culture: The Everyday of Modern Turkey* (pp. 1–21). New Jersey: Rutgers University Press.

Kandiyoti, D. (ed.) (1996) *Gendering the Middle East.* New York: Syracuse University Press.

Kearney, M. (1996) *Reconceptualising the Peasantry: Anthropology in Global Perspective.* Boulder, CO: Westview Press.

Keyder, C. (1993) The genesis of petty commodity production in agriculture – The case of Turkey, in P. Stirling (ed.) *Culture and Economy – Changes in Turkish Villages* (pp. 171–186). Huntingdon: Eothen Press.

Kimber, S., Yang, J., and Cohen, S. (2019) Performing love, prosperity and Chinese hipsterism: Young independent travellers in Pai. *Tourist Studies*, 19 (2): 164–191.

King, B., Dwyer, L., and Prideaux, B. (2006) An evaluation of unethical business practices in Australia's China inbound tourism market. *International Journal of Tourism Research*, 8 (2): 127–142.

Kohn, T. (2018) "Backs" to nature: Musings on tourist selfies, in S. Gmelch and A. Kaul (eds.) *Tourists and Tourism: A Reader* (3rd Edition, pp. 69–77). Long Grove, IL: Waveland Press.

Lash, S., and Urry, J. (1994) *Economies of Signs and Space.* London: Sage.

Latour, B. (1999) On recalling ANT, in J. Law and J. Hassard (eds.) *Actor Network Theory and After* (pp. 15–25). Oxford: Blackwell.

Lee, J. (2022) *Lonely Planet Turkey (Travel Guide)* (16th Edition). Dublin: Lonely Planet.

Leite, N., and Graburn, N. (2009) Anthropological interventions in tourism, in T. Jamal and M. Robinson (eds.) *The Sage Handbook of Tourism Studies* (pp. 35–64). London: Sage.

Little, K. (2020) *On the Nervous Edge of an Impossible Paradise*. New York and London: Berghahn.

Lynch, P. (2017) Mundane welcome: Hospitality as life politics. *Annals of Tourism Research*, 64: 174–184.

MacCannell, D. (1976) *The Tourist: A New Theory of the Leisure Class*. London: Macmillan.

Magliocco, S. (2004) Witchcraft, healing and vernacular magic in Italy, in W. de Blecourt and O. Davis (eds.) *Witchcraft Continued: Popular Magic in Modern Europe* (pp. 151–173). Manchester: Manchester University Press.

Maitland, R. (2021) New urban tourists: In search of the life more ordinary, in A. Condevaux, M. Gravari-Barbas, and S. Guinand (eds.) *Tourism Dynamics in Everyday Places: Before and After Tourism* (pp. 16–22). London: Routledge.

March R. (2008) Towards a conceptualization of unethical marketing practices in tourism: a case-study of Australia's inbound Chinese travel market. *Journal of Travel & Tourism Marketing*, 24 (4): 285–296.

Marcus, J. (1992) *A World of Difference – Islam and Gender Hierarchy in Turkey*. London: Zed.

Massey, D. (1994) *Space, Place and Gender*. Polity Press: Cambridge and Oxford.

Michaud, J. (1991) A social anthropology of tourism in Ladakh, India. *Annals of Tourism Research*, 18: 605–621.

Middelveld, S., van der Duim, R., and Lie, R. (2015) Reef controversies: The case of Wakatobi National Park, Indonesia, in G.T. Jóhannesson, C. Ren, and R. van der Duim (eds.) *Tourism Encounters and Controversies: Ontological Politics of Tourism Development* (pp. 39–52). Burlington, Surrey: Ashgate Publishing.

Milano, C., Cheer, J., and Novelli, M. (2019) Introduction: Overtourism: An evolving phenomenon, in C. Milano, J. Cheer, and M. Novelli (eds.) *Overtourism: Excesses, Discontents and Measures in Travel and Tourism* (pp. 1–17). Wallingford: CABI.

Mills, A. (2007) Gender and Mahalle (neighbourhood) space in Istanbul. *Gender, Place and Culture*, 14 (3): 335–354.

Milne, S., and Ateljevic, I. (2001) Tourism, economic development and the global-local nexus: Theory embracing complexity. *Tourism Geographies*, 3 (4): 369–393.

Molz, J.G., and Gibson, S. (2007) Introduction: Mobilizing and mooring hospitality, in J.G. Molz and S. Gibson (eds.) *Mobilizing Hospitality: The Ethics of Social Relations in a Mobile World* (pp. 1–26). Aldershot: Ashgate.

Molz, J. G., and Paris, C. M. (2015) The social affordances of flashpacking: Exploring the mobility nexus of travel and communication. *Mobilities*, 10 (2): 173–192.

Mottiar, Z., and Tucker, H. (2007) Webs of power: Multiple ownership in tourism destinations. *Current Issues in Tourism*, 10 (4): 279–295.

Munar, A. M., and Gyimóthy, S. (2013) Critical digital tourism studies, in A. M. Munar, S. Gyimóthy and L. Cai (eds.) *Tourism Social Media: Transformations in Identity, Community and Culture* (pp. 245–262). Bingley: Emerald Publishing.

Müftüler-Baç, M. (2012) Gender equality in Turkey, European Parliament Directorate General for Internal Affairs. Retrieved from: https://www.europarl.europa.eu/

RegData/etudes/note/join/2012/462428/IPOL-FEMM_NT(2012)462428_EN.pdf [20/05/2020].

Nash, J. (2007) *Practicing Ethnography in a Globalizing World: An Anthropological Odyssey*. Lanham: AltaMira Press (Rowman & Littlefield Publishers).

Neef, A. (2021) *Tourism, Land Grabs and Displacement*. Routledge: London.

Neumann, M. (1988) Wandering through the museum: Experience and identity in a spectator culture. *Border/Lines*, 15: 19–27.

Notar, B. (2006) *Displacing Desire: Travel and Popular Culture in China*. Honolulu: University of Hawaii Press.

Notar, B.E. (2008) Producing cosmopolitanism at the borderlands: Lonely planeteers and "local" cosmopolitans in Southwest China. *Anthropological Quarterly*, 81 (3): 615–650.

Özbay, C. (2023) Transforming buildings, reorienting lives: The desire for gentrification in Istanbul. *Urban Geography*, 44 (4). doi: 10.1080/02723638.2023.2190265

Ozbay, F. (1995) Changes in women's activities both inside and outside the home, in S. Tekeli (ed.) *Women in Modern Turkish Society: A Reader* (pp. 889–111). London: Zed Books.

Özbudun, E., and Keyman, E.F. (2002) Cultural globalization in Turkey, in P. Berger and S. Huntington (eds.) *Many Globalizations: Cultural Diversity in the Contemporary World* (pp. 296–319). Oxford: Oxford University Press.

Ozenen, A.K. (2018) Arzu's story: Rural Tourism – A necessity for Women's empowerment in Turkey, in S. Cole (ed.) *Gender Equality and Tourism: Beyond Empowerment* (pp. 130–131). Wallingford: CABI.

Özgentürk, J. (2019) Balloon ride consortium under scrutiny in Cappadocia, Hurriyet Daily News (June 9, 2019). Retrieved from: hurriyetdailynews.com/balloon-ride-consortium-under-scrutiny-in-cappadocia-144035 [25/09/2022].

Öztürk, M., Jongerden, J., and Hilton, A. (2018) The (re)production of the new peasantry in Turkey. *Journal of Rural Studies*, 61: 244–254.

Pavlovich, K. (2003) The evolution and transformation of a tourism destination network: The Waitomo Cave, New Zealand. *Tourism Management*, 24: 203–216.

Peristiany, J.G. (1965) *Honor and Shame: The Values of Mediterranean Society*. London, England: Weidenfeld and Nicolson.

Picard, M. (1996) *Bali: Cultural Tourism and Touristic Culture*. Singapore: Archipelago Press.

Ploeg, J.D. (2008) *The New Peasantries: Struggles for Autonomy and Sustainability in an Era of Empire and Globalization*. London: Earthscan.

Pongajarn, C., van der Duim, R., and Peters, K. (2018) Floating markets in Thailand: Same, same, but different. *Journal of Tourism and Cultural Change*, 16 (2): 109–122.

Pratt, M.L. (1992) *Imperial Eyes: Travel Writing and Transculturation*. London: Routledge.

Puwar, N. (2004) *Space Invaders: Race, Gender and Bodies Out of Place*. Oxford: Berg.

Qian, J., and Wei, L. (2020) Development at the edge of difference: Rethinking capital and market relations from Lugu Lake, Southwest China. *Antipode*, 52 (1): 246–269.

Ren, C. (2011) Non-human agency, radical ontology and tourism realities. *Annals of Tourism Research*, 38 (3): 858–881.

Ren, C., Jóhannesson, G.T., and van der Duim, R. (2012) How ANT works, in R. van der Duim, C. Ren, and G.T. Jóhannesson (eds.) *Actor-Network Theory and Tourism: Ordering, Materiality and Multiplicity* (pp. 13–25). London: Routledge.

Rosaldo, R. (1989) Imperialist nostalgia. *Representations*, 26: 107–122.

Rosello, M. (2001) *Postcolonial Hospitality: The Immigrant as Guest.* Stanford, CA: Stanford University Press.

Rowlands, M. (2002) Heritage and cultural property, in V. Buchli (ed.) *The Material Culture Reader* (pp. 105–114). Oxford: Berg.

Roy, A. (2004) Nostalgias of the modern, in N. AlSayyad (ed.) *The End of Tradition?* (pp. 63–86). London: Routledge.

Royce, A. P. and Kemper, R. (2002) Introduction, in R. Kemper and A. P. Royce (eds.) *Chronicling Cultures: Long-Term Field Research in Anthropology* (pp. XIII–XXXVIII). Walnut Creek, CA: AltaMira Press.

Sakızlıoğlu, B. (2018) Rethinking the gender-gentrification nexus, in L. Lees and M. Phillips (eds.) *Handbook of Gentrification Studies* (pp. 205–221). Cheltenham: Edward Elgar Publishing.

Salazar, N. (2010) Tourism and Cosmopolitanism: A view from below. *International Journal of Tourism Anthropology*, 1 (1): 55–69.

Savransky, M., and Stengers, I. (2018) Relearning the art of paying attention: A conversation. *SubStance*, 14 (1): 130–145.

Schiffauer, W. (1993) Peasants without pride, in P. Stirling (ed.) *Culture and Economy – Changes in Turkish Villages* (pp. 65–79). Huntingdon: Eothen Press.

Scott, J. (1976) *The Moral Economy of the Peasant.* New Haven, CT: Yale University Press.

Scott, J. (1995) Sexual and national boundaries in tourism. *Annals of Tourism Research*, 22 (2): 385–403.

Selwyn, T. (1996) *The Tourist Image.* Chichester: Wiley.

Selwyn, T. (2000) An anthropology of hospitality, in C. Lashley and A. Morrison (eds.) *In Search of Hospitality: Theoretical Perspectives and Debates* (pp. 18–37). Oxford: Butterworth-Heinemann.

Sev'er, A. (2005) In the name of fathers: Honor killings and some examples from south-eastern Turkey. *Atlantis: Critical Studies in Gender, Culture & Social Justice*, 30: 129–145.

Sev'er, A., and Yurdakul, G. (2001) Culture of honor, culture of change: A feminist analysis of honor killings in rural Turkey. *Violence against Women*, 7: 964–998.

Simmel, G. ([1910] 1971) Sociability, in D. Levine (ed.) *On Individuality and Social Forms* (pp. 127–40). Chicago: University of Chicago Press.

Simoni, V. (2012) Tourism Materialities: Enacting cigars in touristic Cuba, in R. van der Duim, C. Ren, and G.T. Jóhannesson (eds.) *Actor-Network Theory and Tourism: Ordering, Materiality and Multiplicity* (pp. 59–79). London: Routledge.

Sims, R. (2009) Food, place and authenticity: Local food and the sustainable tourism experience. *Journal of Sustainable Tourism*, 17 (3): 321–336.

Smith, G. (2005) Tourism economy: The global landscape, in C. Cartier and A. Lew (eds.) *Seductions of Place: Geographical Perspectives on Globalization and Touristed Landscapes* (pp. 72–88). London: Routledge.

Smith, S. (2018) Instagram abroad: Performance, consumption and colonial narrative in tourism. *Postcolonial Studies*, 21 (2): 172–191.

Smith, S. (2021) Landscapes for "likes": Capitalizing on travel with Instagram. *Social Semiotics*, 31 (4): 604–624.

Soares, L.E. (1998) Staging the self by performing the other: Global fantasies and the migration of the projective imagination. *Journal for Cultural Research*, 2 (2–3): 288–304.

Sönmez, S. (2001) Tourism behind the veil of Islam: Women and development in the Middle East, in Y. Apostolopoulos, S. Sönmez, and D. Timothy (eds.) *Women as Producers and Consumers of Tourism in Developing Regions* (pp. 113–142). Westport, CT.: Praeger Publishers.

Stirling, P. (1965) *Turkish Village*. London: Weidenfeld and Nicolson.

Swain, M.B. (2009) The cosmopolitan hope of tourism: Critical action and worldmaking vistas. *Tourism Geographies*, 11 (4): 505–525.

Taylor, D.G. (2020) Putting the "self" in selfies: How narcissism, envy and self-promotion motivate sharing of travel photos through social media. *Journal of Travel & Tourism Marketing*, 37 (1): 64–77.

Taylor, J. (1994) *A Dream of England: Landscape, Photography and the Tourist's Imagination*. Manchester and New York: Manchester University Press.

Thomson, R., and McLeod, J. (2015) New frontiers in qualitative longitudinal research: An agenda for research. *International Journal of Social Research Methodology*, 18 (3): 243–250.

Tosun, C. (1998) Roots of unsustainable tourism at the local level: The case of Urgup in Turkey. *Tourism Management*, 19: 595–610.

Tosun, C. (2001) Challenges of sustainable tourism development in the developing world: the case of Turkey. *Tourism Management*, 22(3): 289–303.Tremblay, P. (2019) Cappadocia victim of Turkey ruling party's love of concrete. Al-Monitor. Retrieved from: https://www.al-monitor.com/originals/2019/06/turkeys-cappadocia-under-threat.html#ixzz87OSr1vJ0 [3/3/2023].

Tsing, A. (2005) *Friction: An Ethnography of Global Connection*. Princeton and Oxford: Princeton University Press.

Tsing, A. (2015) *The Mushroom at the End of the World: On Possibility of Life in Capitalist Ruins*. Princeton and Oxford: Princeton University Press.

Tsing, A.L. (2005) *Friction: An Ethnography of Global Connection*. Princeton and Oxford: Princeton University Press.

Tucker, H. (2001) Tourists and troglodytes: Negotiating tradition for sustainable cultural tourism. *Annals of Tourism Research*, 28 (4): 868–891.

Tucker, H. (2002) Welcome to Flintstones-Land: Contesting place and identity in Göreme, Central Turkey, in S. Coleman and M. Crang (eds.) *Tourism: Between Place and Performance* (pp. 143–159). Oxford and New York: Berghahn Books.

Tucker, H. (2003) *Living with Tourism: Negotiating Identities in a Turkish Village*. London: Routledge.

Tucker, H. (2007) Undoing shame: Tourism and women's work in Turkey. *Journal of Tourism and Cultural Change*, 5 (2): 87–105.

Tucker, H. (2009a) Discomfort and shame in tourism encounters: Recognizing emotion and its postcolonial potentialities. *Tourism Geographies*, 11 (4): 444–461.

Tucker, H. (2009b) The cave-homes of Göreme: Performing tourism hospitality in gendered space, in P. Lynch, A. McIntosh, and H. Tucker (eds.) *Commercial Homes in Tourism: An International Perspective* (pp. 127–137). London: Routledge.

Tucker, H. (2010) Peasant-entrepreneurs: A longitudinal ethnography. *Annals of Tourism Research*, 37 (4): 927–946.

Tucker, H. (2011) Success and access to knowledge in the tourist-local encounter: Confrontations with the unexpected in a Turkish community, in D. Theodossopoulos and J. Skinner (eds.) *Great Expectations: Imagination, Anticipation, and Enchantment in Tourism* (pp. 27–39). Oxford: Berghahn Books.

Tucker, H. (2014) Mind the Gap: Opening up spaces of multiple moralities in tourism encounters, in M. Mostafanezhad and K. Hannam (eds.) *Moral Encounters in Tourism* (pp. 199–208). Abingdon: Ashgate.

Tucker, H. (2016a) Empathy and tourism: Limits and possibilities. *Annals of Tourism Research*, 57: 31–43.

Tucker, H. (2016b) Community-based tourism as sustainable development, in I. Egresi (ed.) *Alternative Tourism in Turkey: Role, Potential Development and Sustainabilit* (pp. 335–348). Cham, Switzerland: Springer.

Tucker, H. (2022) Gendering sustainability's contradictions: Between change and continuity. *Journal of Sustainable Tourism*, 30 (7): 1500–1517.

Tucker, H., and Boonabaana, B. (2012) A critical analysis of tourism, gender and poverty reduction. *Journal of Sustainable Tourism*, 20 (3): 437–456.

Urry, J. (1990) *The Tourist Gaze: Leisure and Travel in Contemporary Societies*. London: Sage.

Urry, J. (2002) *The Tourist Gaze: Leisure and Travel in Contemporary Societies* (2nd Edition). London: Sage.

Urry, J., and Larsen, J. (2011) *The Tourist Gaze 3.0*. London: Sage.

Uskul, A.K., and Cross. (2019) The social and cultural psychology of honour: What have we learned from researching honour in Turkey? *European Review of Social Psychology*, 30 (1): 39–73.

Van der Duim, R., Ampumuza, C., and Ahebwa, W.M. (2014) Gorilla tourism in Bwindi Impenetrable National Park, Uganda: An actor-network perspective. *Society and Natural Resources*, 27 (6): 588–601.

Van der Ploeg, J.D. (2008) *The New Peasantries: Struggles for Autonomy and Sustainability in an Era of Empire and Globalization*. London: Earthscan.

Vertovec, S., and R. Cohen. (2002) Introduction: Conceiving cosmopolitanism, in S. Vertovec and R. Cohen (eds.) *Conceiving Cosmopolitanism: Theory, Context and Practice* (pp. 1–22). Oxford: Oxford University Press.

Viken, A. (2014) Destination discourses and the growth paradigm, in A. Viken and B. Granas (eds.) *Tourism Destination Development: Turns and Tactics* (pp. 21–46). Ashgate: Farnham.

Wahnschafft, R. (1982) Formal and informal tourism sectors – A case study in Pattaya, Thailand. *Annals of Tourism Research*, 9: 429–451.

Walsh, N., and Tucker, H. (2010) Tourism 'things': The travelling performance of the backpack. *Tourist Studies*, 9 (3): 223–239.

Wilson, J., and Tallon, A. (2012) Geographies of gentrification and tourism, in J. Wilson (ed.) *The Routledge Handbook of Tourism Geographies* (pp. 103–112). London: Routledge.

Yale, P. (2009) Autumnal hues, *Today's Zaman*, 15 October, 2009.

Yolal, M. (2016) History of tourism development in Turkey, in I. Egresi (ed.) *Alternative Tourism in Turkey: Role, Potential Development and Sustainability* (pp. 23–34). Cham, Switzerland: Springer.

Index

Note: page numbers followed by "n" denote endnotes.

For Product Safety Concerns and Information please contact our EU
representative GPSR@taylorandfrancis.com
Taylor & Francis Verlag GmbH, Kaufingerstraße 24, 80331 München, Germany

www.ingramcontent.com/pod-product-compliance
Lightning Source LLC
Chambersburg PA
CBHW060248220326
41598CB00027B/4021